# THE MATHEMATICS OF FINANCE
## Modeling and Hedging

# The Brooks/Cole Series in Advanced Mathematics

*Paul J. Sally, Jr., Editor*

*The Mathematics of Finance: Modeling and Hedging*
Joseph Stampfli, University of Indiana-Bloomington
Victor Goodman, University of Indiana-Bloomington
© 2001

*Geometry for College Students*
I. Martin Isaacs, University of Wisconsin-Madison
© 2001

*Probability: The Science of Uncertainty with Applications to Investments, Insurance, and Engineering*
Michael A. Bean, University of Michigan-Ann Arbor
© 2001

*A Course in Approximation Theory*
Ward Cheney, The University of Texas at Austin
Will Light, University of Leicester, England
© 2000

*Introduction to Analysis, Fifth Edition*
Edward S. Gaughan, New Mexico State University
© 1998

*Numerical Analysis, Second Edition*
David Kincaid, The University of Texas at Austin
Ward Cheney, The University of Texas at Austin
© 1996

*Advanced Calculus, A Course in Mathematical Analysis*
Patrick M. Fitzpatrick, University of Maryland
© 1994

*Algebra: A Graduate Course*
I. Martin Isaacs, University of Wisconsin—Madison
© 1994

*Fourier Analysis and Its Applications*
Gerald B. Folland, University of Washington
© 1992

# THE MATHEMATICS OF FINANCE
## Modeling and Hedging

**Joseph Stampfli**
**Victor Goodman**

*Indiana University*

BROOKS/COLE
TM
THOMSON LEARNING

Australia • Canada • Mexico • Singapore • Spain • United Kingdom • United States

## BROOKS/COLE

THOMSON LEARNING

Publisher: *Gary Ostedt*
Marketing Representative: *Jay Honeck*
Marketing Team: *Samantha Cabaluna,*
*Beth Kronke, and Karin Sandberg*
Associate Editor: *Carol Benedict*
Editorial Assistant: *Daniel Thiem*
Production Editor: *Mary Vezilich*
Production Service: *Publication Services*

Permissions Editor: *Sue Ewing*
Cover Design: *Vernon T. Boes*
Print Buyer: *Tracy Brown*
Typesetting: *Publication Services*
Cover Printing: *Phoenix Color*
*Corporation*
Printing and Binding: *Maple-Vail*

*For more information about this or any other Brooks/Cole product, contact:*
BROOKS/COLE
511 Forest Lodge Road
Pacific Grove, CA 93950 USA
www.brookscole.com
1-800-423-0563 (Thomson Learning Academic Resource Center)

*For permission to use material from this work, contact us by*
Web:    www.thomsonrights.com
fax:    1-800-730-2215
phone:  1-800-730-2214

Printed in the United States of America

10 9 8 7 6 5 4 3 2 1

**Library of Congress Cataloging-in-Publication Data**

Goodman, Victor, [date–]
   The mathematics of finance: modeling and hedging/Victor Goodman, Joseph Stampfli.
      p. cm. – (The Brooks/Cole series in advanced mathematics)
   Includes index.
   ISBN 0-534-37776-9 (casebound)
   1. Capital market–Mathematical models. 2. Hedging (Finance) I. Stampfli, Joseph G.
   (Joseph Gail), [date–] II. Title. III. Series.

HG4523 .G66 2000
332′.0414–dc21                                                      00-057205

Dedicated to the memory of Harrison "Harry" Roth (1932–1997)
—Joseph Stampfli

*His life was gentle, and the elements*
*So mix'd in him that Nature might stand up*
*And say to all the world, "This was a man."*

To Gail, whom I love
—Victor

# PREFACE

Throughout the nineties, we have seen the synergistic union of mathematics, finance, the computer, and the global economy. Currency markets trade two trillion dollars per day, and sophisticated financial derivatives such as options, swaps, and quantos are commonplace.

Since the appearance of the Black-Scholes formula in 1973, the financial community has embraced an abundant and ever-expanding array of mathematical tools and models. Enrollment in courses presenting these applications of mathematical finance has exploded at schools everywhere. It is driven by the attraction of the material, coupled with enormous employment demand. We expect that the twenty-first century will see even greater growth in these areas, following Kurzweil's law of accelerating returns. The practical analysis of a broad range of market transactions and activities has converted many market devotees to this mode of thinking.

This textbook explains the basic financial and mathematical concepts used in modeling and hedging. Each topic is introduced with the assumption that the reader has had little or no previous exposure to financial matters or to the activities that are common to major equity markets. Exercises and examples illustrate these topics. Often an exercise or example uses real market data.

## To the Instructor

A complete, well-balanced course at the undergraduate level can be based on Chapters 2, 3, 5, 6, 7, 8, and 9. An instructor might touch only briefly on Chapter 1 as an introduction to the financial terminology and to strategies that are employed in trading equity shares. You might wish to return to Chapter 1 repeatedly as you progress through the textbook; the chapter is always there as a convenient reference for market transactions and terminology.

Most undergraduate students seem to be very comfortable with computers, and they appear to pick up the ins and outs of software packages such as Maple™, *Mathematica*™, and Microsoft® Excel very quickly. Each instructor will have to evaluate the proficiency of his or her own students in this area. For example, we have found that Excel is readily available on the Indiana University campus and that students are comfortable in preparing data and reports using this software.

## Acknowledgments

We would like to thank the National Science Foundation for support while preparing some of the material used in this textbook. In particular, we owe a great debt of gratitude to Dan Maki and Bart Ng, principal investigator on the NSF grant, "Mathematics Throughout the Curriculum," for encouraging us to write the book and for their continued support, financial and personal, during the period of creation. We wish to thank our reviewers: Rich Sowers, University of Illinois; William Yin, La Grange College; and John Chadam, University of Pittsburgh.

In November 1999, Joseph Stampfli presented several lectures on financial mathematics at a workshop on this topic in Bangkok, Thailand, sponsored by Mahidol University. We would like to thank the university and, in particular, Professor Yongwemon Lenbury and Ponchai Matangkasombut, then Dean and now President of the university, for their gracious hospitality throughout the visit. It was a truly memorable experience.

We would also like to thank the editorial and production teams at Brooks/Cole for their continuous and timely help. In particular, Gary Ostedt and Carol Benedict did everything an editorial team can do and more. Several unexpected crises arose as the book progressed, and Gary guided us through them with patience, wisdom, and humor. We would also like to thank the other members of the Brooks/Cole team: Mary Vezilich, Production Coordinator; Karin Sandberg, Marketing Manager; Sue Ewing, Permissions Editor; and Samantha Cabaluna, Marketing Communications. We would also like to thank Kris Engberg of Publication Services, who helped us solve hundreds of problems, both large and small; Jerome Colburn, whose contributions as copy editor turned limp doggerel into sparkling prose; and Jason Brown and his production team.

Victor Goodman wishes to thank Devraj Basu for his personal input during the early stages of the manuscript preparation. In addition, Joseph Stampfli would like to thank Jeff Gerlach, a graduate student in Economics at Indiana University. Chapter 11 is entirely due to Jeff's efforts, and he provided solutions to most of the exercises.

## How to Reach Us

Readers are encouraged to bring errors and suggestions to our attention. E-mail is excellent for this purpose. Our addresses are

goodmanv@indiana.edu
stampfli@indiana.edu

A web site for this book is maintained at http://www.indiana.edu/~iubmtc/mathfinance/.

*Victor Goodman*
*Joseph Stampfli*

# CONTENTS

**1  Financial Markets**                                               1

  1.1  Markets and Math                                          1

  1.2  Stocks and Their Derivatives                              2

      1.2.1  Forward Stock Contracts                     3

      1.2.2  Call Options                                7

      1.2.3  Put Options                                 9

      1.2.4  Short Selling                              11

  1.3  Pricing Futures Contracts                                12

  1.4  Bond Markets                                             15

      1.4.1  Rates of Return                            16

      1.4.2  The U.S. Bond Market                       17

      1.4.3  Interest Rates and Forward Interest Rates  18

      1.4.4  Yield Curves                               19

  1.5  Interest Rate Futures                                    20

      1.5.1  Determining the Futures Price               20

      1.5.2  Treasury Bill Futures                      21

  1.6  Foreign Exchange                                         22

      1.6.1  Currency Hedging                           22

      1.6.2  Computing Currency Futures                 23

## 2 Binomial Trees, Replicating Portfolios, and Arbitrage 25

2.1 Three Ways to Price a Derivative 25
2.2 The Game Theory Method 26
    2.2.1 Eliminating Uncertainty 27
    2.2.2 Valuing the Option 27
    2.2.3 Arbitrage 27
    2.2.4 The Game Theory Method—A General Formula 28
2.3 Replicating Portfolios 29
    2.3.1 The Context 30
    2.3.2 A Portfolio Match 30
    2.3.3 Expected Value Pricing Approach 31
    2.3.4 How to Remember the Pricing Probability 32
2.4 The Probabilistic Approach 34
2.5 Risk 36
2.6 Repeated Binomial Trees and Arbitrage 39
2.7 Appendix: Limits of the Arbitrage Method 41

## 3 Tree Models for Stocks and Options 44

3.1 A Stock Model 44
    3.1.1 Recombining Trees 46
    3.1.2 Chaining and Expected Values 46
3.2 Pricing a Call Option with the Tree Model 49
3.3 Pricing an American Option 52
3.4 Pricing an Exotic Option—Knockout Options 55
3.5 Pricing an Exotic Option—Lookback Options 59
3.6 Adjusting the Binomial Tree Model to Real-World Data 61
3.7 Hedging and Pricing the $N$-Period Binomial Model 66

## 4 Using Spreadsheets to Compute Stock and Option Trees 71

4.1 Some Spreadsheet Basics 71
4.2 Computing European Option Trees 74
4.3 Computing American Option Trees 77
4.4 Computing a Barrier Option Tree 79
4.5 Computing $N$-Step Trees 80

# 5 Continuous Models and the Black-Scholes Formula 81

5.1 A Continuous-Time Stock Model 81

5.2 The Discrete Model 82

5.3 An Analysis of the Continuous Model 87

5.4 The Black-Scholes Formula 90

5.5 Derivation of the Black-Scholes Formula 92

    5.5.1 The Related Model 92

    5.5.2 The Expected Value 94

    5.5.3 Two Integrals 94

    5.5.4 Putting the Pieces Together 96

5.6 Put-Call Parity 97

5.7 Trees and Continuous Models 98

    5.7.1 Binomial Probabilities 98

    5.7.2 Approximation with Large Trees 100

    5.7.3 Scaling a Tree to Match a GBM Model 102

5.8 The GBM Stock Price Model—A Cautionary Tale 103

5.9 Appendix: Construction of a Brownian Path 106

# 6 The Analytic Approach to Black-Scholes 109

6.1 Strategy for Obtaining the Differential Equation 110

6.2 Expanding $V(S, t)$ 110

6.3 Expanding and Simplifying $V(S_t, t)$ 111

6.4 Finding a Portfolio 112

6.5 Solving the Black-Scholes Differential Equation 114

    6.5.1 Cash or Nothing Option 114

    6.5.2 Stock-or-Nothing Option 115

    6.5.3 European Call 116

6.6 Options on Futures 116

    6.6.1 Call on a Futures Contract 117

    6.6.2 A PDE for Options on Futures 118

6.7 Appendix: Portfolio Differentials 120

# 7 Hedging 122

7.1 Delta Hedging 122

    7.1.1 Hedging, Dynamic Programming, and a Proof that Black-Scholes Really Works in an Idealized World 123

    7.1.2 Why the Foregoing Argument Does Not Hold in the Real World 124

    7.1.3 Earlier $\Delta$ Hedges 125

7.2    Methods for Hedging a Stock or Portfolio                                     126

    7.2.1    Hedging with Puts                                                126

    7.2.2    Hedging with Collars                                             127

    7.2.3    Hedging with Paired Trades                                       127

    7.2.4    Correlation-Based Hedges                                         127

    7.2.5    Hedging in the Real World                                        128

7.3    Implied Volatility                                                           128

    7.3.1    Computing $\sigma_I$ with Maple                                   128

    7.3.2    The Volatility Smile                                             129

7.4    The Parameters $\Delta$, $\Gamma$, and $\Theta$                              130

    7.4.1    The Role of $\Gamma$                                             131

    7.4.2    A Further Role for $\Delta$, $\Gamma$, $\Theta$                  133

7.5    Derivation of the Delta Hedging Rule                                         134

7.6    Delta Hedging a Stock Purchase                                               135

**8  Bond Models and Interest Rate Options**                                        137

8.1    Interest Rates and Forward Rates                                             137

    8.1.1    Size                                                             138

    8.1.2    The Yield Curve                                                  138

    8.1.3    How Is the Yield Curve Determined?                               139

    8.1.4    Forward Rates                                                    139

8.2    Zero-Coupon Bonds                                                            140

    8.2.1    Forward Rates and ZCBs                                           140

    8.2.2    Computations Based on $Y(t)$ or $P(t)$                           142

8.3    Swaps                                                                        144

    8.3.1    Another Variation on Payments                                    147

    8.3.2    A More Realistic Scenario                                        148

    8.3.3    Models for Bond Prices                                           149

    8.3.4    Arbitrage                                                        150

8.4    Pricing and Hedging a Swap                                                   152

    8.4.1    Arithmetic Interest Rates                                        153

    8.4.2    Geometric Interest Rates                                         155

8.5    Interest Rate Models                                                         157

    8.5.1    Discrete Interest Rate Models                                    158

    8.5.2    Pricing ZCBs from the Interest Rate Model                        162

    8.5.3    The Bond Price Paradox                                           165

    8.5.4    Can the Expected Value Pricing Method Be Arbitraged?             166

    8.5.5    Continuous Models                                                171

    8.5.6    A Bond Price Model                                               171

|  |  |  |  |
|---|---|---|---|
|  | 8.5.7 | A Simple Example | 174 |
|  | 8.5.8 | The Vasicek Model | 178 |
| 8.6 | Bond Price Dynamics |  | 180 |
| 8.7 | A Bond Price Formula |  | 181 |
| 8.8 | Bond Prices, Spot Rates, and HJM |  | 183 |
|  | 8.8.1 | Example: The Hall-White Model | 184 |
| 8.9 | The Derivative Approach to HJM: The HJM Miracle |  | 186 |
| 8.10 | Appendix: Forward Rate Drift |  | 188 |

## 9  Computational Methods for Bonds

190

| 9.1 | Tree Models for Bond Prices |  | 190 |
|---|---|---|---|
|  | 9.1.1 | Fair and Unfair Games | 190 |
|  | 9.1.2 | The Ho-Lee Model | 192 |
| 9.2 | A Binomial Vasicek Model: A Mean Reversion Model |  | 200 |
|  | 9.2.1 | The Base Case | 201 |
|  | 9.2.2 | The General Induction Step | 202 |

## 10  Currency Markets and Foreign Exchange Risks

207

| 10.1 | The Mechanics of Trading |  | 207 |
|---|---|---|---|
| 10.2 | Currency Forwards: Interest Rate Parity |  | 209 |
| 10.3 | Foreign Currency Options |  | 211 |
|  | 10.3.1 | The Garman-Kohlhagen Formula | 211 |
|  | 10.3.2 | Put-Call Parity for Currency Options | 213 |
| 10.4 | Guaranteed Exchange Rates and Quantos |  | 214 |
|  | 10.4.1 | The Bond Hedge | 215 |
|  | 10.4.2 | Pricing the GER Forward on a Stock | 216 |
|  | 10.4.3 | Pricing the GER Put or Call Option | 219 |
| 10.5 | To Hedge or Not to Hedge—and How Much |  | 220 |

## 11  International Political Risk Analysis

221

| 11.1 | Introduction |  | 221 |
|---|---|---|---|
| 11.2 | Types of International Risks |  | 222 |
|  | 11.2.1 | Political Risk | 222 |
|  | 11.2.2 | Managing International Risk | 223 |
|  | 11.2.3 | Diversification | 223 |
|  | 11.2.4 | Political Risk and Export Credit Insurance | 224 |
| 11.3 | Credit Derivatives and the Management of Political Risk |  | 225 |
|  | 11.3.1 | Foreign Currency and Derivatives | 225 |
|  | 11.3.2 | Credit Default Risk and Derivatives | 226 |

| | | |
|---|---|---|
| 11.4 | Pricing International Political Risk | 228 |
| | 11.4.1 The Credit Spread or Risk Premium on Bonds | 229 |
| 11.5 | Two Models for Determining the Risk Premium | 230 |
| | 11.5.1 The Black-Scholes Approach to Pricing Risky Debt | 230 |
| | 11.5.2 An Alternative Approach to Pricing Risky Debt | 234 |
| 11.6 | A Hypothetical Example of the JLT Model | 238 |

## Answers to Selected Exercises 241

## Index 247

# THE MATHEMATICS OF FINANCE
## Modeling and Hedging

# CHAPTER
# 1

# FINANCIAL MARKETS

*If you can look into the seeds of time,*
*And say which grain will grow and which will not,*
*Speak then to me...*

Shakespeare, *Macbeth,* Act I, Scene ii

Note: This chapter is intended to be a glossary. It is designed to introduce concepts, ideas, and definitions as they are needed. We do not recommend that one work through the entire chapter as a unit. Visit it sparingly as needed.

## 1.1 MARKETS AND MATH

Nearly everyone has heard of the New York, London, and Tokyo stock exchanges. Reports of the trading activity in these markets frequently make the front page of newspapers and are often featured on evening television newscasts. There are many other financial markets. Each of these has a character determined by the type of financial objects being exchanged.

The most important markets to be discussed in this book are *stock* markets, *bond* markets, *currency* markets, and *futures and options* markets. These financial terms will be explained later. But first we draw your attention to the fact that every item that is exchanged, or **traded,** on some market is of one of two types.

The traded item may be a **basic equity,** such as a stock, a bond, or a unit of currency. Or the item's value may be indirectly *derived from* the value of some other traded equity. If so, its future price is tied to the price of another equity on a future date. In this case, the item is a **financial derivative;** the equity it refers to is termed the **underlying equity.**

This chapter contains many examples of financial derivatives. Each example will be thoroughly explained in order to make the derivative concept clear to the

1

reader. Our examples will be options based on stocks, bonds, and currencies. Also, we will discuss *futures* and options on futures.

Mathematics enters this subject in a serious way when we try to relate a derivative price to the price of the underlying equity. Mathematically based arguments give surprisingly accurate estimates of these values.

> *The main objective of this book is to explain the process of computing derivative prices in terms of underlying equity prices.*

We also wish to provide the reader with the mathematical tools and techniques to carry out this process. Through developing an understanding of this process, you will gain insights into how derivatives are used, and you will comprehend the risks associated with creating or trading these assets. These insights into derivative trading provide extra knowledge of how modern equity markets work.

The mathematics in this book will emphasize two financial concepts that have had a startling impact over the last two decades on the way the financial industry views derivative trading.

We will emphasize investments that **replicate** equities, and we will explore mathematical models of how equities behave in the **absence of arbitrage opportunities**.

The combination of these two concepts furnishes a powerful tool for finding prices. An example is overdue at this point. In the next section, we present an example in which a *replicating investment* and the *lack of arbitrage opportunities* give us a price for a derivative. This example is worth careful reading.

## 1.2 STOCKS AND THEIR DERIVATIVES

A company that needs to raise money can do so by selling its shares to investors. The company is *owned* by its shareholders. These owners possess **shares** or **equity certificates** and may or may not receive **dividends,** depending on whether the company makes a profit and decides to share this with its owners.

What is the value of the company's stock? Its value reflects the views or predictions of investors about the likely dividend payments, future earnings, and resources that the company will control. These uncertainties are resolved (each trading day) by buyers and sellers of the stock. They exercise their views by trading shares in **auction markets** such as the New York, London, and Tokyo stock exchanges. That is, most of the time a stock's value is judged by what someone else is willing to pay for it on a given day.

What is a stock derivative? It is a specific contract whose value at some future date will depend *entirely* on the stock's future values. The person or firm who formulates this contract and offers it for sale is termed the **writer.** The person or firm who purchases the contract is termed the **holder.** The stock that the contract is based on is termed the **underlying equity.**

What is a derivative worth? The terms of such a contract are crucial in any estimation of its value. As our first example, we choose a derivative with a simple structure so that our main financial concepts, *replicating an equity* and *lack of arbitrage opportunities*, can be easily explained. These will give us a price for this derivative.

We will also explain several trading opportunities in this example that are important for understanding concepts you will encounter later on.

## 1.2.1 Forward Stock Contracts

It is sometimes convenient to have the assurance that, on some specific future date, one will buy a share of stock for a guaranteed price. This *obligation* to buy in the future is known as a

FORWARD CONTRACT

Here are the contract conditions:

- On a specific date, termed the **expiration date,** the holder of this contract **must** pay a prescribed amount of money, the **exercise price,** to the writer of the contract.
- The writer of the contract **must** deliver one share of stock to the holder on the expiration date.

Figure 1.1 is a pictorial view of the exchange of stock and cash in a forward contract, often called a **forward.**

This contract can either be a good deal for the holder on the day of delivery or be a bad one. The outcome depends on the stock price on the expiration date.

### Profit or Loss at Expiration

To state things quantitatively, we will denote the price on that date by $S_T$ and the required exercise price by $X$. The exercise price, $X$, is a known quantity. It is also called the **strike price,** and often the expiration date is referred to as the **strike date.**

The profit or loss to the holder at time $T$ is expressed as

$$S_T - X$$

Is there some way to find a profit or loss price formula that will be useful before the contract expires? The question of what the contract should be worth is not an

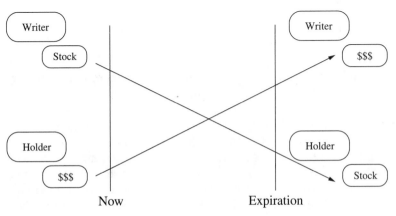

**FIGURE 1.1**
Forward contract

academic one. Modern markets allow a contract holder to sell the contract in the market or purchase new ones during any trading day. In other words, these instruments are **traded.**

You might imagine that the contract has a high value if today's price is much higher than the exercise price, $X$, and expiration is not too far away. On the other hand, if today's stock price is quite low, then perhaps it is nearly worthless. We obtain information about its price by creating an investment that *replicates* some other equity.

### Replicating Investment

Form an investment choice, called a **portfolio,** that consists of one contract (worth $f$) and the following amount of cash:

$$Xe^{-r(T-t)}$$

Now the net worth is

$$f + Xe^{-r(T-t)} \tag{1.1}$$

The exponential factor counteracts interest income.

Any **cash amount** in a portfolio grows by a factor of $e^{r(T-t)}$ in the time from now to expiration. This is reasonable, since the cash will be invested safely. Here, "safely" means that it is placed where the capital will not be at risk from market price changes and can be extracted immediately if needed for some better investment choice. The $r$ denotes the *current* interest rate return on such an investment. During the spring of 2000, an $r$ value of 0.055 per year was used for interest on short-term monetary investments.

The cash amount is adjusted to produce a desired target value for this portfolio on the expiration date of the forward contract.

On that date the portfolio gains a share of stock and pays the exercise price, but the contrived cash amount has grown into exactly the exercise price. In effect, the cash part of the investment disappears and there is no fee.

We can say that on the expiration date this portfolio *replicates* a share of stock. Certainly, the price is correct, since

$$\text{Contract value} + \text{cash amount} = \text{one share of stock}$$

### Trading a Portfolio

Now, here is a surprising aspect of modern markets. Their structure allows this portfolio to be traded *before the expiration date* as though it were an equity. In fact, one can buy this type of contract at any time and set aside some cash; this amounts to *purchasing* one unit of the portfolio.

On the other hand, one usually can sell this contract in a market *even if one does not own it.* An investor serves as the writer of a contract and has the same obligations as a writer. When one sells a basic equity such as a stock without *owning it first,* and then purchases the stock later to make delivery, this activity is referred to as **shorting** or **short selling** the equity. It is possible to sell almost any stock short. Clearly, one can "short" the amount of *cash* in the portfolio just by borrowing some money at the short-term interest rate $r$. In effect, we can *sell* one unit of this portfolio, even if we do not own it to begin with.

We have just explained that one can buy and sell an instrument, the portfolio, that replicates a share of stock on a future date. This leads us to a comparison of prices on an earlier date. We will apply a second major financial principle, *the absence of arbitrage opportunities,* to equate some prices.

### First Arbitrage Opportunity

Suppose that today's price is not consistent with its future value. In fact, let us look at the case when

Contract value + cash amount < one share of stock

This creates a gold mine. The investor can **sell short** quite large amounts of the stock today. This produces instant cash for any investor who is bold enough to sell something that he or she does not own. An investor could use some of the cash to form the correct number of portfolio units to **cover** the short selling.

That is, when the expiration date arrives, the investor neutralizes all the short sales of stock using the **replicating portfolio** value as a stock value. You can see that he or she pockets some cash at the beginning, *regardless of future market behavior,* since it was cheap to cover the short selling.

### Second Arbitrage Opportunity

If the reverse situation holds, that is,

Contract value + cash amount > stock price

the investor could *sell* units of the portfolio *short,* as we explained when we discussed trading the portfolio. Similar arithmetic shows that an investor who covers these short sales with *cheap* stock, purchased immediately after selling short, will still be able to sleep at night.

An investor will not become nervous as the expiration date approaches, because he or she knows that each *short* unit of portfolio will serve only as the liability for exactly one *short* share of stock on this date. Again, some cash is earned at the beginning, *regardless of future market behavior.*

Real markets would not allow either of these money-making schemes to work. We will discuss the reason for this momentarily, but for now, consider the consequence of not having either of the two price inequities discussed above. We obtain the following

### No-Arbitrage Price Equation

Today's contract value + cash amount = today's stock price

We may substitute the values from the net worth equation (1.1) to obtain the formula

$$f + Xe^{-r(T-t)} = S_t$$

The relation can be restated as

$$f = S_t - Xe^{-r(T-t)} \tag{1.2}$$

Equation (1.2) shows that we have "solved" for $f$. To obtain today's price for this contract, one obtains quotes of today's stock price and the short-term interest rate. Once these ingredients are substituted into the formula above, one has a price for a forward stock contract beginning today.

The price should be recomputed each day as the stock price changes and the time to expiration decreases. The $r$ value is unlikely to change in practice, because forward contracts usually expire within 90 days of their initiation. Since this time span is so short, the return on cash invested is far more sensitive to the time to expiration than it is to the rate for one-month or three-month cash investments.

---

**Example**  Suppose we have a forward for Eli Lilly stock that will expire 40 days from now. If the exercise price is $65, and if today's stock price is $64\frac{3}{4}$, what is the contract price today?

We will use an $r$ value of 0.055 per year. The quantities we substitute into equation (1.2) are

$$T - t = 40/365 = 0.1096 \text{ (so that } e^{-r(T-t)} = 0.994)$$

and the two quotes, which are

$$S_t = 64.75 \quad \text{and} \quad X = 65$$

The end result is

$$f = 64.75 - 65(0.994) = 64.75 - 64.61 = \$0.14$$

Another insight this formula gives is that, for the same strike and stock price in the example, a longer-term contract (say, for six months) will have a larger price, because the $e^{-r(T-t)}$ term changes to 0.974. This illustrates the usefulness of equation (1.2). It allows us to compare prices of forward contracts for various expiration dates and various strike prices.

---

## Why Do Arguments Based on *Replication* and *No Arbitrage* Work?

The price formula given by equation (1.2) and the market price of a forward contract cannot be substantially different. Any sizable difference would induce people to follow one of the two investment tricks we discussed. The *guarantee* of profit would induce them to invest enormous sums of money in one of these schemes. Their activities would, in turn, move prices until the *arbitrage opportunity* was driven out of existence by changes in the underlying stock price. For example, if a large amount of a stock is sold short, its present value goes down because so much of it is offered for sale.

Put another way, the market pressures generated by the investment schemes of people who are *certain* to profit would force stock and contract prices into equilibrium values, where the arbitrage opportunity is missing.

## 1.2.2 Call Options

One can purchase an *opportunity* to buy a share of stock in the future for a guaranteed price. This *right, without the obligation*, to buy in the future is known as a

**CALL OPTION**

Here are the option conditions:

- The buyer of the option pays the seller a fee for the option called the premium.
- On the *expiration date,* the holder of this contract **may** pay the writer of the contract the *exercise price.*
- If the writer of the contract receives the exercise price from the holder, the writer **must** deliver one share of stock to the holder on the expiration date.

Figure 1.2 is a pictorial view of the possible exchange of stock and cash. Notice that the holder of this contract does have an investment option. If he or she does not want to buy the share of stock, then the holder avoids paying the exercise price. This obviously happens if the stock price on the expiration date is lower than the exercise price. On the other hand, if the holder sees a high stock price at expiration, then certainly he or she elects to pay the strike price and obtain the valuable share of stock. One says that the option has been **exercised.**

### Profit or Loss at Expiration
In this case, either no exchange occurs or the contract is settled with the writer paying the holder the difference of the stock price and the exercise price. This allows us to describe the possible payoff amount in terms of the stock price at expiration, $S_T$, and the exercise price, $X$. We can say that

$$\text{Call payoff} = \max\{S_T - X, 0\}$$

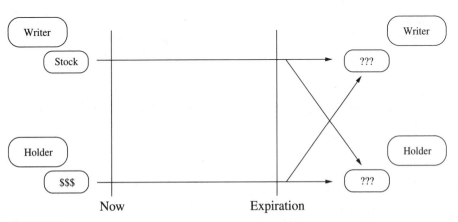

**FIGURE 1.2**
Call option

This "max" formula equals $S_T - X$ if $S_T - X$ is positive; the result is 0 otherwise. We will abbreviate this payoff as

$$(S_T - X)^+$$

The symbol $(x)^+$ denotes $\max\{x, 0\}$.

However, some call options have an even larger payoff. There are two types of calls. We have been discussing a call that has the limitation that the holder may use it only when it expires. This type of call is termed a **European call.**

The other call, an **American call,** is less restricted. The holder is allowed to convert it *any time before expiration.* Of course, once it is exercised, the contract is settled. An American call might have a larger payoff than a European call.

---

**Example—European**   Suppose we hold a call for a share of General Electric (GE), and the call will expire twenty days from now. Suppose the strike price is $88. If today's market price for GE is $84, we might think that the call is worthless since the fee exceeds the current stock price. But it is quite possible that 20 days from now the market price will be higher. Suppose it turns out to be $95$\frac{1}{2}$ on the day of expiration. Then we exercise the option to make a profit of

$$\$95.5 - \$88 = \$7.50$$

A reasonable premium for the option 20 days earlier might have been $4. In this case, the net profit is $3.50. The ratio of profit to investment is spectacular:

$$\frac{3.50}{4.00} = 0.875$$

This is an 87.5% return on the amount we invested in purchasing the call. Unfortunately, it might be the case that GE stock would increase only to $87$\frac{7}{8}$ during the 20 days. Now the call option is worthless and our investment shows a 100% loss. This illustrates the risk and uncertainty associated with buying a call.

---

**Example—Early Exercise**   Suppose we hold an American call for a share of IBM, and the call will expire 15 days from now. Let us suppose the strike price is $105. If today's market price for IBM is $107, we might well wait until the call expires, hoping that 15 days from now the market price will be above $107.

On the other hand, suppose that next week IBM stock jumps to $112 per share. With an American call, we could immediately exercise the option and make a profit of $7 per share before the cost of the option is included. Let us assume we paid $4.50 per call. Our *net* profit on each call would be $2.50. The profit ratio is

$$\frac{2.50}{4.50} = 0.555$$

This is a 55.5% return on the amount we invested in purchasing the call. One might say that this return rate exceeds the higher rate computed in the example for a European call. The reason is that the 55.5% return is achieved over a shorter time span. An American call may produce a large profit in a brief period if the market price of the stock cooperates.

A European option has a restriction on exercise that tends to produce a lower payoff. But at least we have a formula for the payoff in the European case. For this reason, it is easier to estimate the price of this type of option than for an American option.

---

### The Price of a Call

How might we compute the price of a European call option? This is a much more difficult task than the one we faced in Section 1.2.1. In fact, this topic will be deferred to later chapters. The computed price will depend on the mathematical model we wish to use to describe the underlying stock behavior.

Our two financial principles—replication and no arbitrage—do produce a price estimate of a call. The previous section has the following profit or loss formula for a forward contract:

$$S_t - X$$

We can see from the profit formula given in this section that the payoff for a call at expiration is equal to or larger than the payoff just above. In Section 1.2.1 we added some cash to a forward and replicated a stock. So, if we add the same amount of cash to a call, we create an investment that is superior to a stock.

If you now review the first arbitrage opportunity of Section 1.2.1, you will see that lack of arbitrage shows that

$$\text{Call} + Xe^{-r(T-t)} \geq S_t$$

This implies that the call price satisfies the relation

$$\text{Call} \geq S_t - Xe^{-r(T-t)}.$$

This formula, which is our formula price for a forward, is a low estimate for the actual call price.

### 1.2.3 Put Options

One can purchase an *opportunity* to **sell** a share of stock in the future for a guaranteed price, *even if one does not own any stock.* This *right* to sell in the future is known as a

**PUT OPTION**

Here are the option conditions:

- The buyer of the option pays the seller of the option a fee called the premium.
- On the *expiration date,* the holder of this contract **may** give the writer a share of stock or, equivalently, the market price of a share of stock.
- If the writer of the contract receives the share or its price from the holder, the writer **must** pay the exercise fee to the holder on the expiration date.

Figure 1.3 is a pictorial view of the possible exchange of stock and cash.

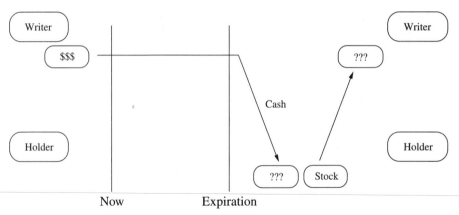

**FIGURE 1.3**
Put option

The holder of this contract does have an investment option. If the holder does not want to give up a share of stock, he or she can negate any exchange. When would this happen? Suppose the stock price on the expiration date is *higher* than the exercise price. This would be a losing exchange for the holder, and the wise investor would reject it.

In contrast, if the holder of a put sees a *low* stock price at expiration, then he or she elects to obtain the fee from the writer and use some of it to obtain a share of stock. The rest is profit.

**Profit or Loss at Expiration**
Nearly always in this case, either no exchange occurs or the contract is settled with the writer paying the holder the difference between the exercise price and the stock price. So we can describe the possible payoff amount in terms of the stock price at expiration, $S_T$, and the exercise price, $X$.

We can say that

$$\text{Put payoff} = \max\{X - ST, 0\}$$
$$= (X - S_T)^+$$

As is the case with call options, many put options have an even larger payoff. We have been discussing a put with the limitation that the holder may use it only when it expires. This type of put is a **European put.**

The other put, an **American put,** is less restricted because the holder can exercise the put *any time before expiration.*

An American put might have a larger payoff than a European put.

---

**Example—Protective Puts** Dr. Brown owns a substantial amount of pharmaceutical stocks that she knew well from her medical work. Merck and Co. is trading at $50 per share, and she believes that the stock price is likely to vary widely in the months ahead.

She wishes to begin selling this stock soon to make some large equipment purchases for her medical practice.

She begins a program of buying puts of approximately three months to expiration, with strike prices set at $45. She has to pay an option premium of $2.80 for each put. This strategy protects her ability to raise cash, since a future low stock price will cause the puts to be exercised. This will allow her to obtain at least $45 for each share called away by a put.

She is guaranteed a minimum price for shares of the stock she sells, as long as she owns puts for these shares. If the stock price stays above the $45 threshold, the puts expire worthless. Her $2.80 expense for each put may be regarded as an "insurance premium" that guarantees the ability to sell some of her stock at or above the level of $45.

---

**Example—Early Exercise** Suppose we hold a put for a share of Ford, and the put will expire fifteen days from now. Suppose the strike price is $35. If today's market price for Ford is $33, we might well wait until the put expires, hoping that fifteen days from now the market price will be *below* $33.

On the other hand, suppose that next week Ford stock falls to $29 per share. With an American call, we could immediately exercise the option and make a profit of $6 per share before the cost of the option is included. Since put options are somewhat inexpensive, our profit ratio on each put might be high.

The fact that a European option can be exercised only at the terminal date is an exercise limitation that tends to produce a lower payoff. As for the case of a call, at least we have a formula for the payoff in the European case. For this reason it is easier to estimate the price of this type of option than an American option.

---

### The Price of a Put

Computing or estimating the price of a put is as difficult as that for a call. This topic will be deferred to later chapters. As in the case of a call price, the computed price will depend on which mathematical model we wish to use to describe the underlying stock behavior.

### 1.2.4   Short Selling

The reader has undoubtedly noticed that the types of *exchanges* allowed in an equity market go far beyond the simple rule that a fixed number of shares can be distributed among those investors interested in purchasing them. The variety of exchanges enhance the availability of equity shares for trading and the availability is termed **liquidity** of the assets traded. Some limitations on liquidity exist.

It is important to understand the following conditions in a short sale.

1. One borrows a concrete, specific number of shares of a stock (usually from a broker) and sells these shares today.
2. The date on which the borrowed shares must be replaced is not specified.
3. If the owner of the borrowed shares decides to sell, then the short seller must borrow other shares and replace the first borrowed shares.

## 1.3   PRICING FUTURES CONTRACTS

A **futures contract** is an agreement between two parties, the **buyer** and the **seller,** to consummate a transaction on a specified date in the future. No goods or money are exchanged today.

---

**Example**  It is January 1, 2000. The buyer and seller enter into an agreement whereby on October 1, 2000, the seller will turn over a barrel of oil to the buyer; and, the buyer will pay the seller the previously agreed-on price of $15.67 for the barrel on that date.

---

There are futures contracts for stock indexes (e.g., Dow, S&P 500, Russell 2000), currencies, interest rates (e.g., U.S. Treasury bond, Eurodollars, LIBOR), and commodities (e.g., wheat, cotton, coffee, copper, crude oil, natural gas), to name a few.

Do futures contracts serve a useful purpose? Yes. Let us look at an example.

---

**Example**  Kellogg's® uses large amounts of corn to produce Kellogg's Corn Flakes®. In March 2001, it might purchase a futures contract on corn for December 2001 (after harvest of the 2001 crop). This way it is sure of obtaining a "reasonable" price. It avoids the possibility of a drought driving corn prices to astronomical levels. Similarly, the farmer or seller of the contract ensures that she gets a "reasonable" price even if there is an enormous crop and prices fall. Both are happy. Neither Kellogg's nor the farmer wants to be in the weather prediction business. Note that the contract smooths out the price and avoids uncertainty and possible disaster for both parties.

---

One can easily make up similar examples for airlines and jet fuel prices and for large multinational corporations and currency exchange rates. We now come to the question of pricing futures contracts. For stock futures, we will be able to specify an exact price with an argument that is crucial to your understanding of financial derivatives.

### Stock Futures

Suppose a buyer agrees to take delivery of a share of stock at a date denoted by $T$, and the buyer and seller wish to determine a price of $\$X$ that the buyer should pay the seller *when the buyer takes the delivery.* What amount, $X$, would a buyer and seller agree on?

Let us approach the question in the following way. We have agreed to deliver one share of stock at time $T$. At that point the stock price may be much higher or lower than today's price. If we wait until time $T$, we are at the mercy of the market. Is there some way we can protect ourselves against the extremes of market prices? Stop here and think for a minute; you should be able to discover a simple solution.

*Simple Solution.*  We just buy one share of the stock today (at $t = 0$ and price $S_0$) and put it in the drawer. We hold it and deliver at time $t = T$. To pay for the stock, we borrow $S_0$ today at the riskless rate $r$. At delivery we repay the loan, which at time $T$ has grown to $S_0 e^{rT}$. To break even and protect ourselves, we should charge $S_0 e^{rT}$ for the futures contract.

So far we have phrased the pricing procedure in terms of protection (or survival). Let us look at the transaction from a slightly different perspective. This new point of view *forces* the price of the future to $S_0 e^{rT}$. Suppose someone offers to pay us $S_U > S_0 e^{rT}$ for the future. We accept with glee and alacrity. We follow the strategy described above. At settlement (time $T$) we end up with a net of

$$S_U - S_0 e^{rT} > 0$$

(Our customer pays us $S_U$, and we repay the $S_0 e^{rT}$ we owe.) We have made a surething, riskless profit. Under these circumstances, the person buying the future would discover hundreds of sellers eager to do business, and he or she would lower the offered price $S_U$ eventually to $S_0 e^{rT}$ (plus perhaps a transaction cost).

So far we have seen that we should expect to see $S_U \le S_0 e^{rT}$. Can we reverse this argument to show that $S_U \ge S_0 e^{rT}$? There is an asymmetry in futures contracts, and the answer is yes for stocks but no for commodities such as oil, soybeans, copper, and hemp. Let us take them one at a time.

Suppose I am offered a futures contract for one share of stock at time $T$ for a price $S_D < S_0 e^{rT}$. I acquire the contract at $S_D$ (no money changes hands) and sell short one share of stock at the present price $S_0$. I invest the $S_0$ at the riskless rate $r$. At time $T$ my invested funds have grown to

$$S_0 e^{rT}$$

At this point I unwind the position by paying $S_D$ to the futures seller, take my share of stock, and give it to the short seller to close out the transaction. I have a riskless profit of

$$S_0 e^{rT} - S_D$$

Thus, sellers of stock futures will be besieged by offers from buyers if their price falls below $S_0 e^{rT}$. Thus, the futures price is driven to $S_0 e^{rT}$.

### Futures Price for Stocks

We can present a progression of the *futures* price for a fixed delivery date, $T$, in terms of the daily or hourly market prices, $S_t$, for the stock. The formula just shown applies, but we acknowledge that the time to expiration, $T - t$, continually decreases. Then, the progression of futures prices, denoted by $F_t$, is related to the stock price by the equation

$$F_t = S_t e^{r(T-t)} \tag{1.3}$$

**Remark.** Suppose a seller of futures had determined that a certain company, say Electrotech, was in extreme difficulty and sold futures for $S_D < S_0 e^{rT}$. You buy some and follow the strategy described above. As anticipated, Electrotech falls below $S_D$ to, say, $S_\tau$. You profit; namely, you are ahead $S_0 e^{rT} - S_D$. The seller profits; she is ahead $S_D - S_\tau$. Who loses? The short seller owning the stock that you borrowed and sold.

TABLE 1.1
**Soybeans (CBT)—5,000 bu.; cents per bu.**

|  | Open | High | Low | Close | Change | Contract High | Contract Low | Vol. |
|---|---|---|---|---|---|---|---|---|
| July | 619 | 622.5 | 617.25 | 621.5 | + 2.25 | 753 | 611 | 56,563 |
| Aug | 610 | 613.5 | 609 | 613 | + 3.5 | 745 | 603 | 19,275 |
| Sept | 596.25 | 597.25 | 593 | 597 | + 4.25 | 723 | 587 | 6,529 |
| Nov | 587 | 592.5 | 585.5 | 591.75 | + 5.25 | 717 | 580.5 | 48,282 |
| Ja99 | 595.5 | 599 | 594.5 | 598.75 | + 6 | 701.5 | 588 | 4,234 |
| Mar | 602 | 604.5 | 601.5 | 604.5 | + 4.75 | 694 | 595 | 1,389 |
| May | 608 | 608 | 606 | 608 | + 4 | 671 | 600 | 408 |
| July | 612.5 | 615 | 610 | 614.5 | + 5.5 | 728 | 604 | 1,026 |

Est. vol. 44,000, vol. Mon 51,097; open int. 139,005, +923.


TABLE 1.2
**Coffee (CSCE)–37,500 lbs.; cents per lb.**

|  | Open | High | Low | Close | Change | Contract High | Contract Low | Vol. |
|---|---|---|---|---|---|---|---|---|
| July | 124.25 | 125.75 | 122.30 | 124.85 | + 0.70 | 191.00 | 120.00 | 15,752 |
| Sept | 123.00 | 124.60 | 122.00 | 124.05 | + 0.60 | 186.00 | 122.00 | 8,162 |
| Dec | 120.00 | 121.50 | 119.00 | 120.50 | + 0.30 | 157.50 | 119.00 | 7,172 |
| Mr99 | 116.00 | 117.50 | 116.00 | 116.50 | + 0.25 | 154.00 | 116.00 | 2,919 |
| May | 114.70 | 115.25 | 114.00 | 114.50 | + 0.25 | 151.00 | 114.00 | 1,162 |
| July | 113.25 | 113.50 | 112.75 | 112.50 | + 0.25 | 131.00 | 112.75 | 728 |
| Sept | 112.00 | 112.00 | 111.00 | 111.00 | + 0.205 | 123.00 | 111.00 | 688 |

Est. vol. 10,326 vol. Mon 11,740; open int. 36,583, +1,580.


*Commodities.* As can be seen from Tables 1.1. and 1.2, the future prices for soybeans and coffee do not follow the $S_0 e^{rT}$ rule. Commodities are different from stocks in two important respects:

1. You cannot sell short a bushel of wheat.
2. Unlike stocks, there is added supply in the form of the current crop coming into the market.

   Note from Table 1.1, which covered the 1998–1999 season, that the soybean prices fall until the January 1999 ("Ja99") contract, where they turn up. A very large crop was anticipated in 1998, and the market was saying "you will be able to buy all the beans you want at prices well below the present spot price of $6.19$\frac{1}{2}$."
3. There are storage costs for commodities, but they should drive the later contract prices up, not down.

   We present an example using stock index futures.

**Example** The following are closing prices on June 8, 1998 (*The Wall Street Journal,* June 9, 1998, p. C18).

| | |
|---|---|
| S&P 500 | 1115.72 |
| Sept S&P 500 Index future | 1129.2 |
| 13-week Treasury bill rate | 4.995% |
| Time Period | 102 days |

Since

$$e^{0.0487425} = 1.049995 \quad \text{(annualized yield)}$$

our projected value for the S&P 500 Index future is

$$1115.72e^{0.0487425 \times 102/365} = 1131.02$$

## 1.4 BOND MARKETS

At the present time, the U.S. bond market represents about 800 billion dollars (*Barron's,* June 12, 2000, p. MW77; Federal Reserve Data Bank), whereas the U.S. stock market has a combined market value of about 15 trillion dollars (*The Wall Street Journal,* June 8, 2000, p. C1).

In its basic form a bond is a loan. It reflects a promise by a borrower, the *seller* of the bond, to repay the amount borrowed at a specific time, plus interest at an agreed-upon rate. Note that the word "promise" is embedded in the term *promissory note.* The bond market is the channel through which governments and corporations that need to borrow money are matched with investors who have funds to lend. The bond market provides monetary liquidity that is vital to an economy's health.

Bond dealers at security firms act as intermediaries, buying from *issuers* and selling to *buyers.* This activity forms the **primary market** for bonds. Bond dealers also maintain an active **secondary market** in bonds, bidding for bonds, already issued, that investors wish to sell. Dealers also offer bonds, already issued, from their inventory to investors who wish to buy. It is possible to sell bonds short, just as in the case of stock. Bonds are bought by institutions such as insurance companies, mutual funds, pension funds, and financial institutions as well as by individual investors. They are also actively traded, and the average holding period for the 30-year Treasury bond is two weeks.

Every bond has a **face** or **par value,** which is the sum that the buyer of the bond will receive when the bond matures. When a bond is sold as a part of a new issue, its price is fixed.

### BONDS ARE OF TWO MAJOR TYPES—DISCOUNT AND COUPON.

A *discount,* or **zero-coupon,** bond pays the owner only the face value of the bond at the time of maturity. *Coupon bonds* pay the face value at maturity, and they also generate fixed, periodic payments known as **coupons** over the lifetime of the bond.

For example, U.S. Treasury bonds have *face values* in multiples of 1000 dollars. Their *maturity period* can range from 2 years to 30 years. Typically the two-year bond is a *coupon bond,* whereas the 30-year Treasury bond is a *zero-coupon bond.*

When the bond is first sold as part of a new issue, its price is fixed by the issuer. From then on, its *secondary market* price goes up or down, partly in response to the general level of interest rates.

Interest rates change in response to a number of things: changes in demand and supply of credit, fiscal policy, Federal Reserve policy, exchange rates, economic conditions, and, most important for the bond market, changes in perceptions regarding inflation. By inflation, we mean a **persistent rise in prices of basic goods.** This price pattern lowers the value of bonds as it reduces the future buying power of the fixed payments from the bond.

The bond market often reacts negatively to positive economic news because of the fear of rising inflation. Conversely, *negative news,* such as higher unemployment, reduces inflationary concerns and tends to *raise* bond prices. When interest rates go up, the price of a bond goes down because its coupon rate becomes less desirable than the rates of newly issued bonds of similar quality. If interest rates fall, the bond's coupon rate becomes more attractive to investors, which drives up the price.

## 1.4.1 Rates of Return

When we buy a bond at a market price different from the bond's face value, three numbers are commonly used to measure the annual rate of return we get on our investment:

**Coupon rate:** the periodic payment measured annually, as a percentage of the bond's face value

**Current yield:** the annual payout as a percentage of the current market price of the bond

**Yield-to-maturity:** the percentage rate of return paid on the bond if it is bought and held to its maturity date

*Yield-to-maturity* is perhaps the best measure of the return rate. It is obtained by *discounting.*

### Discounting a Future Payment

A promise of a payment of $1000 one year from now, no matter how sincerely made, is not worth $1000 today. A lesser amount today could be invested safely and would grow to reach $1000 after one year.

To be precise, the value, $V$, of this promise today could be earning *some yearly return rate, R,* as some investment. We can express the investment growth to the $1000 payoff as $Ve^R = 1000$.

---

**Example** Suppose we know we can invest money, with almost no risk, at the rate of 6% per year. Then

$$Ve^{0.06} = 1000$$
$$V(1.062) = 1000$$

so that

$$V = \$948$$

---

This division by the growth factor is termed **discounting.** If the payoff will occur at some future time $T$ and the present time is $t$, then the growth equation reads $Ve^{(T-t)R} = P$, where $P$ is the payoff size. If we know the investment rate $R$, we obtain $V$ as

$$V = e^{-R(T-t)}P \tag{1.4}$$

For a bond, **yield-to-maturity** is *the rate R* that, when used to discount all future payouts from the bond to their present value, will yield the present price of the bond. In other words, equation (1.4) is solved for $R$ once we know the present market value $V$.

Let us solve equation (1.4) in the case of a zero-coupon bond. If the bond has face value 1 and matures at time $T$, then we can compute $R$ in terms of its current price, $P(t, T)$, as

$$R(t) = \frac{-\ln P(t, T)}{T - t} \tag{1.5}$$

There is no easy way to calculate the yield-to-maturity of a coupon bond. Some comparison rules for the three return rates are the following:

- The yield-to-maturity is always greater than the current yield, which in turn is always greater than the coupon rate, if a bond happens to be selling below its face value.
- The opposite is true for a bond selling at a price higher than its face value.
- A bond selling at its face value has its coupon rate equal to its current yield and yield-to-maturity.

## 1.4.2 The U.S. Bond Market

The yield-to-maturity on the 30-year Treasury bond is widely regarded as the bell-wether rate—the leader—of the bond market. It mirrors the direction of interest rates.

**Treasury bills (T-bills)** are debt instruments with maturities of one year or less. They are backed by the full faith and credit of the U.S. government. Treasury bills are low-risk investments with a broad and liquid secondary market. Both Treasury bonds and bills are purchased by investors at Treasury auctions. T-bills with maturity periods of 13 and 26 weeks are auctioned each Monday, and the 52-week T-bill is auctioned every four weeks on Thursday. Treasury bonds are auctioned less frequently. The 30-year Treasury bonds, for example, are auctioned early in February, August, and November.

Security firms (underwriters) act as middlemen in the sale and dissemination of bonds and notes. The ability of securities firms to price securities effectively and to underwrite issues of government and corporate debt depends on their ability to price their instruments accurately. These firms serve as underwriters and market makers; these activities require the firms to buy and sell the security. Underwriters and market makers are essential to the smooth functioning of the debt markets. Highly liquid markets in bonds require readily available funding sources.

**Repurchase agreements (repos)** are the most important source of liquidity in the Treasury markets. In a typical repo agreement, a securities dealer wishing to finance a bond position sells the bonds to a cash investor while simultaneously agreeing to repurchase them at a later date at an agreed-upon price. The bond is placed with the buyer as collateral. A reverse repo is the opposite agreement, with the buyer agreeing to buy bonds and the seller agreeing to repurchase them at a future time. The term of the repo can be custom-tailored for any period, ranging from overnight to a year.

The repo market originated as a means by which securities dealers could finance their bond positions, and it still serves that purpose today. The outstanding volume of repos is enormous.

### 1.4.3   Interest Rates and Forward Interest Rates

The **$n$-year interest rate** is the interest rate on an investment that is made for a period of time *starting today* and lasting for $n$ years. It is the yield on a discount or zero-coupon bond that matures in $n$ years. It is also referred to as the $n$-year zero-coupon yield. We can find its value from equation (1.5) by obtaining a quote for the zero-coupon bond price in the secondary market.

**Forward interest rates** are the rates of interest implied by current interest rates for periods of time *beginning in the future.*

---

**Example**  Suppose the one-year interest rate is 8% and the two-year rate is 8.5%. The *forward rate for the second year* is the interest rate that, when combined with 8% for year one, gives 8.5% for the two-year rate. Since we assume that interest is continuously compounded, this rate is calculated using exponentials.

**The Method:**  We will denote the forward rate by $f(1, 2)$. We invest $100. It grows to $\$100e^{0.08}$ at the end of the first year. If we terminate this investment but reinvest the money for the second year, the final amount would be

$$\$e^{f(1,2)}100e^{0.08} = \$100e^{0.08+f(1,2)}$$

But the two-year rate is 8.5%. That is, the same $100 grows to

$$100e^{2\times0.085}$$

To rule out arbitrage opportunities, these amounts must match. We equate them:

$$100e^{2 \times 0.085} = 100e^{0.08 + f(1,2)}$$

so that

$$0.17 = 0.08 + f(1, 2)$$

We see that the forward rate, $f(1, 2)$, is 9%.

In general, if $r_1$ is the $T_1$-year rate of interest and $r_2$ is the $T_2$-year rate for a longer maturity, then the **forward interest rate** between $T_1$ and $T_2$, is given by

$$f(T_1, T_2) = \frac{r_2 T_2 - r_1 T_1}{T_2 - T_1} \qquad (1.6)$$

### 1.4.4  Yield Curves

The zero-coupon **yield curve** is the graph of zero-coupon yields plotted against their maturity in years. There is a corresponding curve showing the rate for coupon-paying bonds plotted against their maturity. *The Wall Street Journal* publishes the Treasury yield curve daily. An example is shown in Figure 1.4.

While a yield curve is typically higher for longer maturities, the *zero-coupon* yield curve will always be higher than the yield-curve for coupon-paying bonds. A coupon-paying bond produces payoffs before the maturity dates of the bond. These payments are discounted less than the final payoff and produce a lower yield.

Equation (1.6) may be rewritten as

$$f(T, T_2) = s + (s - r)\frac{T}{T_2 - T}$$

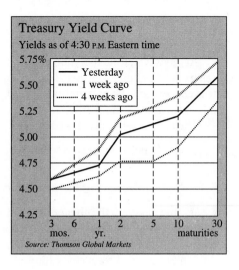

**FIGURE 1.4**
U.S. Treasury bond yield curve for March 11, 1999 (Source: *Wall Street Journal,* March 11, 1999)

This shows that if $s > r$, then $f > s > r$, so that forward rates are higher than current yields. In that case if we let $T_2$ approach $T$, so that $s$ approaches $r$, we see that the forward rate, for a very short period of time, beginning at time $T$ is

$$f(T, T) = r + T \frac{\partial r}{\partial T}$$

This is known as the *instantaneous forward rate* for maturity $T$.

## 1.5   INTEREST RATE FUTURES

The most popular interest rate future is the **Treasury bond interest rate future** traded on the Chicago Board of Trade (CBOT). In this contract, the buyer is guaranteed to receive a government bond with more than 15 years to maturity on the first day of delivery. In fact, any Treasury bond that cannot be exchanged by the government for cash within 15 years from the first day of delivery can be delivered to satisfy this contract.

Treasury bond prices are quoted in dollars and 32nds of a dollar. The quoted price is for a bond with a face value of 100 dollars. A quote of 91-05 means that the indicated price for a bond with a face value of 100,000 dollars is $100,000 \times 91\frac{5}{32}/100 = 91,156.25$ dollars. The quoted price is not the same as the cash price paid by the purchaser. The relationship between the cash price and the quoted price is

(Cash price) = (quoted price) + (accrued interest since last coupon date)

Treasury bond futures prices are quoted in the same way as the Treasury bond prices themselves. Delivery can take place at any time during the delivery month.

The party with the short position who delivers the bond receives a payment defined by the conversion factor of the bond. The conversion factor for a bond is equal to the value of the bond on the first day of the delivery month. The cash received by the party with the short position is equal to

(Quoted futures price) $\times$ (conversion factor) + (accrued interest on bond)

The accrued interest is that on each $100 of face value.

At any given time there are about 30 bonds that can be delivered on the CBOT Treasury futures bond contract. These vary widely as far as coupon rate and maturity date is concerned. The party with the short position can choose which of the available bonds is cheapest to deliver.

The cost of purchasing a bond for delivery is

(Quoted price) + (accrued interest since last coupon date)

### 1.5.1   Determining the Futures Price

If we assume that both the cheapest-to-deliver bond and the delivery date are known, the Treasury bond futures contract is a futures contract on a security providing the holder with known income. We can determine the price $F_t$ of the futures contract as we did in Section 1.4 for a stock future.

Let $C$ denote the present value of all the coupon payments on the bond. We know the amount and the time each coupon will be paid, so that we can discount each payment to obtain its present value. Let $P$ be the current price of the bond, $T$ the time when the futures contract matures, and $t$ the present time.

We will see that

$$F_t = (P - C)e^{r(T-t)}$$

where $r$ is the riskless rate applicable between $T$ and $t$. The expression $P - C$ appears because of the following comparison of investment choices:

**Portfolio A** consists of one *futures contract,* worth $f$ at time $t$ on the bond, plus an amount of cash, $Xe^{-r(T-t)}$.

**Portfolio B** consists of one *bond* plus borrowings of amount $C$ at the riskless rate.

The coupon payments from the bond can be used to repay the borrowings so that portfolio B has the same value as one unit of the bond at time $T$. In portfolio A, the cash invested at the riskless rate will grow to $X$ at time $T$, and will be used to pay for the bond at time $T$. Both portfolios consist of one bond at time $T$. The principle of no arbitrage implies that they must have the same value at $t$.

$$f + Xe^{-r(T-t)} = P - C$$

or

$$f = P - C - Xe^{-r(T-t)}$$

The futures price of the contract $F$ is the value of $X$ that makes $f = 0$.

Hence we see that

$$F = (P - C)e^{r(T-t)}$$

### 1.5.2 Treasury Bill Futures

Treasury bills have maturities of a year or less. The Treasury bill of shortest duration is the 90-day Treasury bill. In **the Treasury bill futures contract,** the underlying asset is a 90-day Treasury bill. For example, if the futures contract matures in 180 days, the underlying asset is a 270-day Treasury bill. These contracts may be regarded as contracts dependent on the short rate. Let us try to calculate the price of such a contract.

Suppose the futures contract matures in $T$ years and the underlying asset matures in $T + \frac{90}{360} = \frac{1}{4}$ years $= T_1$ years. Let $r$ and $T_1$ denote the rates for riskless investments maturing in $T$ and $T_1$ years. Assume that the T-bill underlying the futures contract has a face value of 100 dollars. Its current value $V_0$ is given by

$$V_0 = 100e^{-r_1 T_1}$$

This is the current value of our asset. Since the contract matures at time $T$ and the asset pays no dividends, we see that the futures price $F$ is $e^{rT}$ times the asset price.

Hence

$$F = e^{rT} V_0 = 100e^{rT - r_1 T_1}$$

Recall from (1.6) that the forward rate between $T$ and $T_1$, $f$, is related to $r$ and $r_1$ via the equation

$$f(T, T_0) = \frac{rT - r_1 T_1}{T - T_1}$$

Hence, we may rewrite the futures price $F$ as

$$F = 100e^{f(T,T_1)(T - T_1)}$$

This equation shows that the price of the futures contract will be the same as the price of the T-bill if the 90-day interest rate on the day of delivery is the same as the current forward rate for that date.

## 1.6 FOREIGN EXCHANGE

Forward contracts and options are widely used to manage foreign exchange risks. These risks arise when companies enter into transactions involving payments in a foreign currency. For example, a German company may sell its products to a U.S. company and receive payments in U.S. dollars 30 days hence. It will have to convert the dollars into German marks. The amount of marks it receives depends on the prevailing U.S. dollar/German mark exchange rate. The company has taken on a risky position, even though it may be receiving a known amount in U.S. dollars.

We illustrate the use of forward contracts and options to manage exchange rate risk and to achieve guaranteed cash flows. This process is known as *hedging*.

### 1.6.1 Currency Hedging

Suppose Ger Beta Hans (GBH), a German corporation, is due to receive one million U.S. dollars six months from now. It will have to convert this into German marks. If it remains unhedged, the corporation will benefit if the U.S. dollar/German mark rate goes up, since it will receive more marks for the dollars. Similarly, it will lose if the rate goes down.

The simplest thing to do to reduce this risk is to sell a forward contract based on the dollar/mark exchange rate. Suppose the six-month forward rate is 1.4300. That is, 1.43 marks equals one dollar. If GBH sells a forward contract worth one million U.S. dollars, it is guaranteed to receive 1,430,000 marks against delivery of the 1 million U.S. dollars in six months.

If the exchange rate fell over the next six months, GBH would be covered, but if the rate went up, GBH would not benefit, beacuse it would be locked into the forward rate of 1.430 marks per dollar. The forward contract buys GBH the certainty of obtaining 1.43 marks per dollar and has no up-front costs or premiums. The contract is inflexible, however, and does not allow GBH to participate in the appreciation of the dollar/mark exchange rate.

As an alternative to a forward contract, GBH may choose to buy a U.S. dollar put/mark call. The underlying equity for the option would be the exchange rate. For example, GBH might choose to buy a U.S. dollar put option, with a strike equal to the current forward exchange rate, namely 1.4300 marks per dollar. The option gives GBH the right to sell the dollars that it receives in six months and receive 1.43 marks for every dollar it sells.

This option would guarantee GBH 1.43 marks per dollar—protection if the exchange rate fell. If the exchange rate rose above 1.43, this option would expire worthless; the option gives GBH the right, not the obligation, to sell the dollars. It could then sell the dollars at the prevailing rate and would receive more than 1.43 marks per dollar. The put option thus allows GBH to participate in any appreciation of the exchange rate.

This put option does have a premium. The six-month U.S. dollar put with a strike of 1.43 would cost 0.0466 marks per U.S. dollar. Converting this into U.S. dollars and multiplying by one million, we see that this is approximately $32,600. The break-even rate for this strategy is $1.4300 - 0.0466 = 1.3844$ marks per dollar.

If GBH knew that the exchange rate would be less than 1.3844 in six months, then it would come out ahead of the "do nothing" approach by buying the put option. Moreover, if the rate ends up above $1.4300 + 0.0466 = 1.4766$, GBH is better off with the put strategy rather than the purchase of a forward contract. It will fully benefit from the appreciation of the rate, whereas with the forward contract it would have given this opportunity away. The put strategy performs best when there are large swings.

GBH could purchase an option with a lower strike rate. This would provide protection against a larger move in the rate, but it would cost less. The purchase of a put option thus requires an up-front payment as opposed to a forward. The purchase of a put might thus be regarded as the purchase of an asset. It provides real and tangible benefits for GBH, but it carries a price tag. The forward contract guarantees a certain rate, and hence does not represent an asset or a liability.

### 1.6.2 Computing Currency Futures

Consider the following futures contract involving U.S. dollars and German marks. At time $T$, A will deliver one mark to B. B in turn will pay A a certain number of dollars at that time. What is the fair price for the futures contract? What should A charge B for one mark at time $T$?

| | |
|---|---|
| Exchange rate today | one dollar $= \alpha$ marks |
| Risk-free rate, U.S. | $= r_d$ |
| Risk-free rate, Germany | $= r_m$ |

We construct the following diagram:

$$
\begin{array}{ccc}
 & \text{Dollars} & \text{Marks} \\
T = 0 & 1 \longleftrightarrow \alpha \\
 & \downarrow & \downarrow \\
\text{Time } T & e^{r_d T} \cdot 1 \longleftrightarrow e^{r_m T} \alpha
\end{array}
$$

We reason as follows. We have \$1 and we can pursue two courses of action:

1. We can invest the \$1 at the U.S. rate, in which case it is worth $e^{r_d T}$ dollars at time $T$, or
2. We can convert the \$1 to $\alpha$ marks and invest it at the German rate. In this case, it is worth $e^{r_m T} \alpha$ marks at time $T$.

We can pursue either course, so they must have equivalent values at time 0. (Their values at time $T$ could be very different.) Thus,

$$
\begin{array}{ccc}
e^{r_d T} & = & e^{r_m T} \alpha \\
\text{Dollars} & & \text{Marks} \quad \text{Time 0}
\end{array}
$$

Hence, one mark at time $T$ is worth

$$ e^{(r_d - r_m)T} / \alpha $$

dollars today, and that is the fair price for the futures contract. As a practical matter, to implement the trade, A would borrow $e^{-r_m T} / \alpha$ dollars. He would convert it to $e^{-r_m T}$ marks today. By time $T$, it would have grown to one mark. At time $T$, he must repay

$$ (e^{-r_m T} / \alpha) e^{r_d T} = e^{(r_d - r_m)T} / \alpha $$

dollars, which we have decided is the fair price. Note that this trade is riskless from A's perspective. It is attractive from B's perspective because there will be no "surprises" at time $T$.

# CHAPTER 2

# BINOMIAL TREES, REPLICATING PORTFOLIOS, AND ARBITRAGE

*First place your army so that you cannot lose.*

Sun Tzu, *The Art of War*

## 2.1 THREE WAYS TO PRICE A DERIVATIVE

As we stated in Section 1.2.2, a (European) call option on a stock is the right, but not the obligation, to purchase the stock on a certain day, termed the strike date or date of expiration. The purchase price is agreed on in advance; it is termed the strike price or exercise price. A (European) put option on a stock is a similar right, an option to *sell* the stock on a certain date, as described in Section 1.2.3. It is understood that one does not need to *own* a share of stock to exercise a put.

Table 2.1 shows a typical range of options traded on a particular day for a particular stock. The price of an option appears to go down as the expiration date approaches and to be higher the farther the strike price is from the current price. What should the price of the option be?

---

**Example—A Call Option** We have a stock presently priced at $100. In exactly one year the stock price will be either $90 or $120. We are not given probabilities. The current interest rate is 5% (a dollar invested today is worth $1.05 in one year). What is a fair price for the option on the stock with a strike price of $105 expiring in one year?

---

**TABLE 2.1**
**Ford Option Trades, Nov 15, 1997 (current stock price $28\frac{5}{8}$)**

| Strike price | Expiration date | Vol. | Call price |
|---|---|---|---|
| 27 1/2 | Dec | 10 | 1 9/16 |
| 27 1/2 | Mar | . . . | . . . |
| 30 | Nov | 1059 | 5/16 |
| 30 | Dec | 84 | 3/4 |
| 30 | Mar | 2100 | 1 5/8 |
| 32 1/2 | Dec | 59 | 1/4 |
| 32 1/2 | Mar | 504 | 11/16 |
| 35 | Mar | 1055 | 5/16 |

Source: *The Wall Street Journal*, November 16, 1997.

We will present *three* methods that can be used to answer this question. One method can be called the game theory approach, the second the replicating-portfolio method, and the third the probabilistic or expected-value approach. Some readers may feel we do not have enough information to answer the question. They are right. We will have to make some assumptions. The assumptions are all part of the approach or solution to the problem, and we feel no compunction at adding them as we go along. The first assumption, common to all three approaches, is that the stock price must be one of only two specified values at the expiration date.

## 2.2 THE GAME THEORY METHOD

Let $V$ = price of the option, and let $S$ = price of the stock. We will construct a portfolio as follows: We buy $a$ shares of the option and $b$ shares of the stock. The numbers $a$ or $b$ may be negative. If $b$, for example, is negative, it indicates that we are short-selling the stock. Let $\Pi_0$ = the value of the portfolio at time $t = 0$.

$$\Pi_0 = aV + bS$$

At this point we do not know $a$ or $b$.

Next we will value the portfolio at time $t = 1$. Since, according to our assumption, there are two possible states (scenarios) at $t = 1$, we take them separately.

$$(\text{Up state, } S = \$120) \qquad \Pi_1 = a(120 - 105) + b120$$

Why does the term $120 - 105$ appear? We convert the option into stock and sell the stock. The cost of the conversion is $105 per share, and the sale price is $120 per share.

$$(\text{Down state, } S = \$90) \qquad \Pi_1 = a \cdot 0 + b90$$

We seem to have made little progress but now we can introduce a brilliant stroke. We cannot control the stock performance, so we use a trick from game theory.

### 2.2.1 Eliminating Uncertainty

We will choose $a$ and $b$ so that $\Pi_1$ *does not* depend on the outcome. Thus, we set

$$a(120 - 105) + b120 = a \cdot 0 + b90$$

This leads to the equation $15a = -30b$, and we make the choice $b = +1$ and $a = -2$. This strategy tells us we should sell *two* options and buy *one* share of stock. If we do so, the outcome is deterministic; we can ignore the real world.

### 2.2.2 Valuing the Option

Note that

$$\Pi_0 = -2V + 1 \cdot 100$$

and

$$\Pi_1 = -15 \cdot 2 + (+1)120 = 90$$

We could have taken the funds $\Pi_0$ and invested them at 5%. So the value of that investment should equal the value $\Pi_1$ of the stock-option trade.

In other words,

$$1.05\Pi_0 = 1.05(100 - 2V) = \Pi_1 = 90$$

Thus, $100 - 2V = \frac{90}{1.05}$, and so $V = 7.14$.

### 2.2.3 Arbitrage

Suppose a dealer was willing to buy (or sell) the option at \$7.25. Our pricing method tells us that this is too much. So, we do the following. We buy one share of stock and sell two options.

Our cost for this position is

$$100 - (7.25) \cdot 2 = \$85.50$$

We borrow the \$85.50 for one year at the short rate

$$r = 0.04879 \text{ (so that the yearly rate is } e^r = 1.05)$$

We are long one share of stock, are short two options, and owe \$85.50.

At the end of one year we close out the position. The stock–option combination produces a net \$90. We owe $\$85.50 \cdot 1.05 = \$89.775$ on the loan. We are left with a profit (riskless) of

$$\$90 - 89.775 = \$0.225$$

This may not seem like much, but note that this is a riskless transaction. In real life, one might purchase one million shares of stock and sell two million options.

Suppose the dealer was offering the options at \$7.00. They are now under-priced; so we reverse the procedure. We buy two options and sell one share of stock short. Our net cash position is now

$$\$100 - 2 \cdot 7 = \$86$$

We invest the \$86 at the short rate, $r$, mentioned earlier. We are long two options, short one share of stock, and have invested \$86 safely. At the end of the year we unwind the position. Our \$86 is now worth

$$86 \cdot 1.05 = \$90.30$$

We close out the option–stock position at a cost of \$90. We are left with a riskless profit of

$$\$90.30 - 90 = \$0.30$$

The foregoing calculations tell us something very important. If the option price is out of line (in this universe), then investors or arbitrageurs will take advantage of the opportunity for riskless profit and trade options and stocks at huge volume. These actions will affect prices, as discussed in Section 1.2.1. Thus, the arbitrageurs will drive the price back into line. That is, they will force the price to converge to \$7.14. Markets respond so quickly that we can assume that any arbitrage that could be done has already been done and no opportunities for arbitrage remain, as described in Chapter 1.

In the real world, stock prices do not follow a binomial model, and arbitrage is still possible in some situations.

### 2.2.4 Game Theory Method—A General Formula

We now repeat the foregoing argument to obtain a general expression for our derivative value. We assume that our stock can take on only two values at time $\tau$. If the stock is in the up state, $S_u$, then the derivative has the value $U$; if the stock is in the down state, $S_d$, then the derivative has the value $D$ (see Figure 2.1).

We construct a portfolio by buying one derivative at price $V_0$ and selling $a$ shares of stock. The initial portfolio value is

$$\Pi_0 = V_0 - aS_0$$

We choose $a$ so that the value of the portfolio is independent of the final state of the stock. Since

$$\text{Up}: \qquad \Pi_u = U - aS_u$$

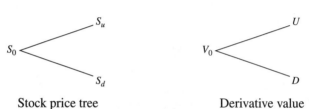

Stock price tree　　　　　　　　　　Derivative value

**FIGURE 2.1**
Trees for stock price and derivative

and

$$\text{Down}: \quad \Pi_d = D - aS_d$$

we require that

$$U - aS_u = D - aS_d$$

Thus, we choose

$$a = \frac{U - D}{S_u - S_d} = \frac{\Delta V}{\Delta S}$$

This ratio, $\Delta V / \Delta S$, plays an extremely important role in the pricing of options and derivatives. We will meet it again and again. We incorporate $a$ into our calculation:

$$\text{Initial cost of portfolio} = V_0 - aS_0$$

$$\text{Final value of portfolio} = U - aS_u$$

Since this portfolio investment carries no risk and the riskless return rate is $r$, we must have

$$V_0 - aS_0 = e^{-r\tau}(U - aS_u)$$

We solve this equation to obtain the **derivative price formula**

$$V_0 = aS_0 + (U - aS_u)e^{-r\tau} \tag{2.1}$$

Equation (2.1) gives the correct price for the derivative, since, if $V_0$ did not match the price, there would be an arbitrage opportunity to enjoy a riskless profit. We leave it to the reader to supply the details.

## EXERCISES

1. A unit of stock is valued at \$110. In one year the stock price will be either \$130 or \$100. Suppose that the corresponding derivative value will be $U = \$10$ or $D = \$0$. The current one-year riskless interest rate is 4% ($e^{r\tau} = 1.04$). Find the derivative price at $t = 0$.

2. Suppose that a stock has the same values as in Exercise 1 and the interest rate is 4%. Suppose that the derivative is a call option whose exercise price is \$100. Find the call option price at $t = 0$.

3. A unit of stock is valued at \$100. In one year the stock price will be either \$130 or \$90. Suppose that the corresponding derivative value will be $U = \$0$ or $D = \$5$. The current one-year interest rate is 5%. Find the derivative price at $t = 0$. This derivative is a put option with an exercise price of \$95.

4. A unit of stock is valued at \$60. In one year the stock price will be either \$75 or \$50. Suppose that the derivative is a put option whose exercise price is \$60. The current one-year interest rate is 5%. Find the put option price at $t = 0$.

## 2.3   REPLICATING PORTFOLIOS

We encountered a powerful technique in Section 2.2 in which two investments were combined to produce a fixed return. We will illustrate that other investment combinations can imitate financial derivatives.

**FIGURE 2.2**

### 2.3.1   The Context

As before, we have a stock where $S_0$ is the stock price at $t = 0$ and the stock can achieve one of two possible values at time $t = \tau$ (see Figure 2.2).

We also have a derivative security $V$, whose value at $t = \tau$ will depend on the performance of $S$. If $S$ goes up, $V$ will be worth $U$. If $S$ goes down, $V$ will be worth $D$. What is a fair price for $V$ today?

Although (2.1) gives a price for $V$, this value was obtained indirectly, as a by-product of an investment whose return is fixed. We can obtain more information about derivative values by employing an ingenious device known as a *replicating portfolio*.

### 2.3.2   A Portfolio Match

We need to fix one more item. The riskless interest rate will be denoted by $r$, and we make use of the short-term lending and borrowing possible at this rate. Let us record such amounts of money in terms of a **bond** by assuming the bond is initially worth $1. Then the value of such a bond at time $t$ is $e^{rt}$. Our portfolio, $\Pi$, will consist of $a$ units of the stock and $b$ units of the bond. The value, $\Pi_0$, of the portfolio at time $t = 0$ is just

$$\Pi_0 = aS_0 + b$$

Let us compute the value of $\Pi$ at $t = \tau$. Our stock model gives two future values for the portfolio:

$$\text{Up state} \qquad \Pi_\tau = aS_u + be^{r\tau}$$

$$\text{Down state} \qquad \Pi_\tau = aS_d + be^{r\tau}$$

We now set

$$aS_u + be^{r\tau} = U \qquad\qquad (2.2)$$
$$aS_d + be^{r\tau} = D$$

so that the value of our portfolio, $\Pi$, is **identical** to the value of the derivative security. The portfolio **replicates** $V$. Since the portfolio and the derivative have the same value at $t = \tau$, they should have the same value today. After all, they are indistinguishable at the next future date. We conclude that

$$V_0 = aS_0 + b$$

This expression for $V_0$ has a striking form once we solve for $a$ and $b$ using the equations (2.2). These are two linear equations whose solution is

$$a = \frac{U - D}{S_u - S_d}$$

$$b = \left[U - \frac{U - D}{S_u - S_d}S_u\right]e^{-r\tau} \tag{2.3}$$

Although these expressions look complicated, they produce simpler expressions for portfolio values. Combining the last three expressions, we find that

$$V_0 = aS_0 + (U - aS_u)e^{-r\tau}$$

which is just our earlier pricing formula (2.1). We focus on the expression for $V_0$ (which is just $aS_0 + b$).

$$V_0 = \frac{U - D}{S_u - S_d}S_0 + \left[U - \frac{U - D}{S_u - S_d}S_u\right]e^{-r\tau}$$

and we separate the $U$ terms and the $D$ terms to get

$$V_0 = U\left[\frac{S_0}{S_u - S_d} + e^{-r\tau} - \frac{S_u}{S_u - S_d}e^{-r\tau}\right] + D\left[\frac{-S_0}{S_u - S_d} + \frac{S_u}{S_u - S_d}e^{-r\tau}\right]$$

$$= e^{-r\tau}U\left[\frac{e^{r\tau}S_0}{S_u - S_d} - \frac{S_d}{S_u - S_d}\right] + e^{-r\tau}D\left[\frac{S_u}{S_u - S_d} - \frac{e^{r\tau}S_0}{S_u - S_d}\right]$$

There is something very special here. The $U$ coefficient, ignoring the exponential term, is

$$q = \frac{e^{r\tau}S_0 - S_d}{S_u - S_d}$$

and the $D$ coefficient,

$$\frac{S_u - e^{r\tau}S_0}{S_u - S_d}$$

is just $1 - q$. So our portfolio value does simplify to

$$V_0 = e^{-r\tau}\left[qU + (1 - q)D\right] \tag{2.4}$$

Note the similarities between equations (2.4) and (2.1). Our results are consistent with those of Section 2.2.4. However, we are employing an investment mixture of stock and cash that mimics the behavior of a derivative. This approach allows equation (2.4) to apply to several time steps, as we will see in Section 2.6.

If we use the numbers from the example in Section 2.2 and we calculate $V_0$ using equation (2.4), the result is still $7.14.

### 2.3.3 Expected Value Pricing Approach

Equation (2.4) states that the present value of a portfolio is obtained by discounting ($e^{-r\tau}$ is referred to as a discount factor) an average of future portfolio values. Indeed, the formula for $q$ would serve as some sort of probability if we could verify the condition $0 \leq q \leq 1$. Let us look at the value for $q$ once more:

$$q = \frac{e^{r\tau}S_0 - S_d}{S_u - S_d} \tag{2.5}$$

If $q$ were negative, this stock would be a great buy. The numerator in (2.5) is negative, so

$$e^{r\tau}S_0 < S_d$$

The stock's worst future value, $S_d$, would exceed the return we would get by initially investing \$$S_0$ in the bond. Note that the bond return would be \$$e^{r\tau}S_0$. This is another example of the sure money-making scheme known as **arbitrage**, and we believe this is too good to be true in the real world. Equally unrealistic is the case where $1 - q$ is negative. We see from the expression just before equation (2.4) that

$$1 - q = \frac{S_u - e^{r\tau}S_0}{S_u - S_d}$$

and, if the numerator were negative, the stock would be a dud. In this case, the best future value of the stock, $S_u$, is not as large as the return on an initial bond investment of \$$S_0$. There would be no reason to buy such a stock. Again, a real market would not support such stock behavior.

So, it is realistic to assume that $q$ is a probability, and this motivates us to rewrite our simple portfolio value in equation (2.4) as

$$\begin{aligned} V_0 &= e^{-r\tau}\left[\, qU + (1-q)D \,\right] \\ &= e^{-r\tau}E_q\left[V_1\right] \end{aligned} \tag{2.6}$$

The subscript in equation (2.6) indicates that we are using the specially computed *no-arbitrage pricing* probability, $q$, given by (2.5). These probabilities are also referred to as *risk-neutral probabilities*.

### 2.3.4   How to Remember the Pricing Probability

A good way to remember the $q$-value for expected value pricing is to consider the simplest, one-period investment: buy a single share of stock. This corresponds to the tree in Figure 2.3.

We may set up the expected-value equation (2.6) even though we already know the $V_0$ value, which is just $S_0$. According to the equation, we must have

$$e^{r\tau}S_0 = qS_u + (1-q)S_d \tag{2.7}$$

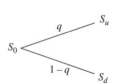

**FIGURE 2.3**

Although (2.7) is just one expected-value example, it is the most important one for understanding the role of $q$. Take a few moments to *solve this equation* for $q$. The formula for $q$ matches the one in (2.5). You will appreciate the fact that this tree enables one to calculate the pricing probability once we are given three stock values and an interest rate.

---

**Example** A stock has the current value $50. After one year, its value will be either $55 or $40. The one-year interest rate is 4%.

Suppose we wish to know the price of two calls, one with an exercise price of 48 and the other with an exercise price of 53. Also, we wish to price a put whose exercise price is $45.

How do we use equation (2.6) to find these three prices quickly?

***Step 1.*** Obtain $q$ from the stock tree (Figure 2.4).

Equation (2.7) states that

$$1.04 \cdot 50 = q55 + (1 - q)40$$

From

$$52 = q55 + (1 - q)40$$

we obtain

$$12 = q55 - 40q = 15q$$

so $q = \frac{12}{15} = 0.8$.

***Step 2.*** Average the $U$ and $D$ derivative values.

1. If a call has an exercise price of 48, then $U = 7$ and $D = 0$. The call price is

$$\frac{1}{1.04}\left[q \cdot 7 + (1 - q) \cdot 0\right] = \frac{5.6}{1.04} \doteq \$5.38$$

2. If the exercise price is instead $53, then $U = 2$. The call price is

$$\frac{1}{1.04}\left[0.8 \cdot 2 + 0\right] = \frac{1.6}{1.04} \doteq \$1.54$$

3. If a put option has an exercise price of 45, then $U = 0$ and $D = 5$. The put price is

$$\frac{1}{1.04}\left[0 + 0.2 \cdot 5\right] = \frac{1.0}{1.04} \doteq \$0.96$$

---

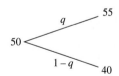

**FIGURE 2.4**

It will be important to remember one other quantity in addition to the $q$ value. In this section we have focused on a portfolio that matches or *replicates* the results of some other equity. The key idea is to hold the correct number of shares of the stock. Our formula for this number of shares, equation (2.3), is just

$$\text{Number of shares} = \frac{U - D}{S_u - S_d} \qquad (2.8)$$

This quantity is precisely the $a$ that appeared in the game theory method (2.2.4).

Notice that this ratio compares the change in the equity *being replicated* to the change in the stock price. This ratio is termed the **delta quantity** when it determines an investment process. You will see the pricing probability, $q$, and the delta quantity, $\Delta$, appear in many calculations of portfolio behavior later on.

## 2.4   THE PROBABILISTIC APPROACH

Let us begin by thinking about this situation using one characteristic of real markets. We know the stock price is $100, the up price is $120, and the down price is $90. Suppose we were viewing real market behavior over a one-year period. A reasonable choice of $p$, the probability that the stock will rise (see Figure 2.5), is one where the expected return of the stock is on the order of 15%. This return is much larger than if we invested the $100 in a secure bank account, and the following calculations suggest why this is so.

A $p$ value that roughly matches this expected return is $p = 0.90$. This seems to produce an attractive situation. The expected payoff is given by

$$E(P) = 0.9(120 - 100) + 0.1(90 - 100) = \$17$$

Notice that the *expected* return each year is 17%. But there is still some uncertainty. Since only a *probability of success* is involved, we find that

90% of the time you make $20

and

10% of the time you lose $10

Many investors would purchase stock under these circumstances. The healthy price increase of the stock would offset the occasional loss, so this investment choice is attractive to those who can afford the risk of some loss.

However, each investor is different. How do we decide what is a reasonable amount of risk and reward for this stock? To circumvent this impossible task we introduce a hypothetical investor, hereafter designated by H.I., who has the following

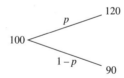

**FIGURE 2.5**
Stock price tree

characteristics:

1. The H.I. is risk neutral; this is the opposite of "a bird in the hand" outlook. A risk-neutral investor is risk indifferent; a certain dollar is no more preferable than an expected dollar. Most people are not risk neutral. The insurance industry is based on this fact.
2. Our H.I. thus has no preference between the stock introduced above and a riskless investment.

Given these assumptions, what value of $p$ in our stock model shown in Figure 2.5 will make the stock 5% return (0.05) equally attractive to the investor?

If we form a portfolio $\Pi$ consisting of one share of stock, then $\Pi_0 = \$100$ and

$$E[\Pi_1] = p120 + (1 - p)90$$
$$= 30p + 90$$

after one year. If we had simply invested the $100 at the riskless rate, then the value of that selection would be $105 at the end of one year. Being risk neutral, our H.I. sees these investments as equal. That is,

$$30p + 90 = 105$$

which implies that

$$p = 0.5$$

***Important Caveat.*** The value we have just found does not necessarily correspond to investor sentiment or to some real but unknown probability associated with the stock. It is simply the probability that produces a stock return equivalent (in the hypothetical value sense) to the riskless return.

Now, let us take this $p$ value and use it to compute the expected value for a call option on the stock. Let $C$ be our call price, and recall that the exercise price is $105. Then

$$E[C] = p(120 - 105)^+ + (1 - p)(90 - 105)^+$$
$$= 0.5 \cdot 15 + 0.5 \cdot 0 = \$7.50$$

However, we pay for the call today and receive our payoff in one year, so we should discount the expected return on the call. We then arrive at the value

$$E[C]/1.05 = \$7.50/1.05 = \$7.14$$

Astonishingly, this is exactly the price we calculated using the game theory and replicating-portfolio methods.

## EXERCISES

1. A unit of stock is valued at $90. In one year the stock price will be either $105 or $80 (no probabilities are given). The current (time 0) riskless interest rate is 4%. Use the game

theory approach of Section (2.2.4) to price a put option on the stock, where the option expires in one year with a strike price of $110.

2. A unit of stock is valued at $110. In one year the stock price will be either $102 or $122 (no probabilities are given). The current (time 0) riskless interest rate is 4%. Use the expected-value equation (2.6) to price a put option on the stock, where the option expires in one year with a strike price of $110.

3. Using the methods and solutions of Sections (2.2.4) and (2.3.2), show that the game theory and the expected-value approaches to option pricing are equivalent.

**Hint:** Begin with the game theory approach for a general option, with values $U$ and $D$. Use the game theory answer for $V$, and show that your formula has the form of a discounted expected value.

4. (a) Consider the call option price computed by the game theory method in Exercise 1. Show by numerical example that (i) a higher call price would create an arbitrage opportunity; (ii) a lower call price would create an arbitrage opportunity.
   (b) Consider the put option price computed by the game theory method in Exercise 3(b). Show by numerical example that (i) a higher put price would create an arbitrage opportunity; (ii) a lower put price would create an arbitrage opportunity.

5. Suppose that X is a constant and that a derivative has one of the future values $U = S_u - X$ or $D = S_d - X$. Use the expected-value equation to show that

$$V_0 = S_0 - e^{-r\tau}X$$

6. A unit of stock is valued at $100. In one year, the stock price will be either $115 or $90. The current riskless interest rate is 4%.

   (a) Use the expected-value equation (2.6) to price a put option on the stock that expires in 1 year with a strike price of $95.

   (b) Suppose another derivative has one of the future values $U = 0$ and $D = $15$. Explain why $V_0$ must be 3× the price computed in part (a).

   More exercise on this material can be found in the Chapter Review Exercises.

## 2.5 RISK

Risk plays a central role in the investment world. Risk can be bought, sold, or packaged. In general, option dealers prefer to limit or minimize their risk. Let us see how they can do this in the binomial world. We will illustrate the method with an example. The dealer will *hedge*, the position, or hedge the risk.

---

**Example** Our stock, Flash Inc., is selling at $60. Its future one year price performance is contained in the binomial model shown in Figure 2.6.

**FIGURE 2.6**
Binomial model for Flash, Inc.          Stock price                    Call option price

The dealer wishes to offer a call option with a strike of $65 and an expiration of one year. The riskless rate is 0.048. The input values for equation (2.1) are

$$S_0 = 60 \qquad U = 15 \qquad r = 0.048$$
$$S_u = 80 \qquad D = 0 \qquad \tau = 1$$
$$S_d = 50$$

First, we calculate $V_0$.

$$a = \frac{U - D}{S_u - S_d} = \frac{15}{30} = \frac{1}{2}$$

$$V_0 = aS_0 + (U - aS_u)e^{-r\tau}$$

$$= \frac{1}{2}(60) + \left(15 - \frac{1}{2}(80)\right)e^{-0.048}$$

$$= \$6.16$$

The dealer offers to sell the call at $6.35 and buy it at $6.00. The difference between $6.00 and $6.35 is called *the dealer's spread*.

A customer purchases 100,000 shares of the call option (1000 calls in the jargon of the option market) at $6.35 per share. The dealer now has a very risky position. She decides to hedge the risk by buying stock. How much should she buy?

***Answer:*** She should buy $\Delta \cdot 100,000$ shares of stock, where

$$\Delta = \frac{U - D}{S_u - S_d} = \frac{1}{2}$$

(This calculation should look familiar.) So she buys 50,000 shares of stock at a cost of $3,000,000. Note that she received

$$\$6.35 \cdot 100,000 = \$635,000$$

for the calls. This forced her to borrow $2,375,000 at 0.048 in order to purchase the stock. Is the dealer protected against the slings and arrows of uncertainty? Let us hop ahead and see.

***Case 1. Stock moves to 80.*** The dealer's stock is worth $80 \cdot 50,000 = \$4,000,000$. She owes $15 \cdot 100,000 = \$1,500,000$ to redeem the calls and $2,375,000 \cdot e^{0.048} = \$2,493,750$ to redeem the loan. Thus, her net final position is

$$4,000,000 - (1,500,000 + 2,493,750) = \$6,250$$

So she enjoys a profit of $6,250 in this case.

***Case 2. Stock drops to 50.*** The dealer's stock is now worth

$$50 \cdot 50,000 = \$2,500,000$$

The calls have expired worthless, so she has no liability there. But she must redeem the loan, which comes to

$$2,375,000 \cdot e^{0.048} = \$2,493,750$$

Thus, her net final position is

$$2,500,000 - 2,493,750 = \$6,250$$

Hence, she achieves a profit of $6,250 in both cases.

The astute reader will notice that we could have taken a shortcut in Case 2, since

$$2,500,000 = 4,000,000 - 1,500,000$$

and the loan payment is the same in both cases.

The hedging technique just exhibited is called a *delta hedge*. We will encounter it again in Chapters 3 and 5. Delta hedging is used hundreds of times a day in the real world of Wall Street. It is a standard tool for reducing risk in the options arena.

## EXERCISES

1. You are an options dealer. Given a binomial model for stock prices, you sell call options where

| $S_0$ | $S_u$ | $S_d$ | $X$ | $r$ | $\tau$ | Number of shares $N$ |
|---|---|---|---|---|---|---|
| 50 | 60 | 40 | 55 | 0.05 | $\frac{1}{2}$ | 1000 |

   (a) Calculate the fair market price for the call option.
   (b) Assume you sell 1000 shares of the option for the fair market price $+\$0.10$. How many shares of stock should you buy to hedge the sale?
   (c) What is your profit, independent of the outcome of the stock price?

2. Assume the same conditions as in exercise 1, except that you, as a dealer, sell 5,000 puts with a strike of 50 and an expiration of three months ($\tau = \frac{1}{4}$).
   (a) Calculate the fair market price for the put.
   (b) How many shares of stock should you use to hedge the sale?

   *Hint:* Caution: $\Delta$ will be negative, which means you will sell short a certain number of shares of stock.

   (c) What is your profit, if you sell the put for the fair market price $+\$0.12$?

3. This exercise is patterned on Exercises 1 and 2. You are given the following data:

| | Type | $S_0$ | $S_u$ | $S_d$ | $X$ | $r$ | $\tau$ | Number of shares, $N$ |
|---|---|---|---|---|---|---|---|---|
| (a) | Call | 80 | 90 | 75 | 80 | 0.048 | 1 | 2,000 |
| (b) | Put | 80 | 90 | 70 | 75 | 0.05 | $\frac{1}{2}$ | 1,000 |
| (c) | Call | 70 | 80 | 50 | 55 | 0.046 | $\frac{1}{4}$ | 2,000 |
| (d) | Put | 40 | 45 | 30 | 40 | 0.05 | $\frac{1}{2}$ | 6,000 |
| (e) | Call | 64 | 72 | 60 | 66 | 0.047 | $\frac{1}{3}$ | 4,000 |
| (f) | Put | 24 | 30 | 20 | 22 | 0.05 | 1 | 3,000 |

   (a) Calculate the fair market price of the put or call.
   (b) Find the number of shares of stock to buy or sell to hedge the position.
   (c) Assume you sell the put or call for the fair market price $+\$0.10$. Find your profit, independent of stock performance.

## 2.6 REPEATED BINOMIAL TREES AND ARBITRAGE

In the next chapter we look at repeated binomial trees in detail, but let us look at the two-step case before moving on. Our model is a stock whose price moves twice, and each move is to a simple up or down price, as in the preceding sections.

Step 1 consists of a stock price movement from $S_0$ to either of $S_u$ or $S_d$. However, we will specialize these values to

$$S_u = u \cdot S_0 \qquad \text{and} \qquad S_d = d \cdot S_0$$

where two fixed parameters, $u > 1$ and $d < 1$ have been chosen.

Step 2 then consists of a similar stock price movement from either of these two values. The stock may increase by a factor $u$ or decrease by the factor $d$. Figure 2.7 summarizes the four possible end results of these two steps.

Since $u \cdot d = d \cdot u$, we can simplify our tree by representing it as shown in Figure 2.8.

Now suppose that a financial derivative has a specific price for each of the end results in the stock tree. The three possible values will be $U$, $M$, or $D$, corresponding to the stock values $u^2 S_0$, $ud S_0$, or $d^2 S_0$. These results are illustrated in Figure 2.9.

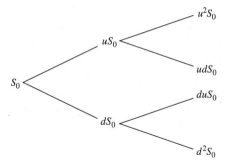

**FIGURE 2.7**
Two-step binomial model

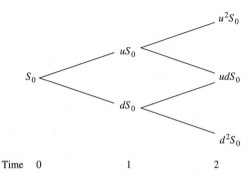

**FIGURE 2.8**
Simplified stock tree

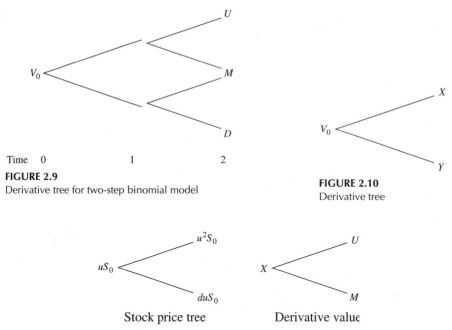

Time    0                    1                    2

**FIGURE 2.9**
Derivative tree for two-step binomial model

**FIGURE 2.10**
Derivative tree

**FIGURE 2.11**
Partial stock and derivative trees in
two-step binomial model

Stock price tree          Derivative value

How do we arrive at a price $V_0$ for this derivative? The key is to find possible prices for the derivative at time $t = 1$. Suppose we had these prices, $X$ and $Y$. Then we are faced with the tree problem shown in Figure 2.10. But we can solve this using any of the methods in sections 2.2 through 2.4. Equation (2.1) shows that

$$V_0 = aS_0 + (X - aS_u)e^{-r\tau}$$

where

$$a = \frac{\Delta V}{\Delta S} = \frac{(X - Y)}{(uS_0 - dS_0)}$$

Clearly, every quantity, including $V_0$, is determined once we know $X$ and $Y$.

Here is a method for finding $X$. Consider the portions of the stock and derivative trees corresponding to the prices $uS_0$ and $X$ and the price changes shown in Figure 2.11. Here again, the stock moves either up or down from its initial value of $uS_0$, and the two possible derivative prices corresponding to each stock value are known. We use equation (2.1) to compute the $X$ value that is the fair price of the derivative in this scenario.

The computation of $Y$ is carried out for a similar pair of trees, when the initial stock price is $dS_0$ and the two possible derivative values are $M$ and $D$. To summarize, we computed intermediate prices and used these values to obtain $V_0$. We will develop systematic ways of completing these steps in the next chapter.

## 2.7  APPENDIX: LIMITS OF THE ARBITRAGE METHOD

We have just seen that the replicating portfolio method, combined with the absence of arbitrage, exactly determines an option price. It converts what appears to be a stochastic situation into a deterministic one—*in the binomial context*. However, a stock that takes one of only two possible prices is not very realistic. Can we extend this approach to a stock that takes on three values as shown in Figure 2.12? The answer is no (unless we make unreasonable assumptions). Let us see why.

Our portfolio will (as before) consist of

$a$      units of the stock $S$

$b$      units of a bond (price is \$1 at an interest rate $r$)

the value of the portfolio at $t = 0$ is

$$\Pi_0 = a \cdot S + b$$

We now set the portfolio value at $t = 1$ equal to the option value for the three scenarios:

| | |
|---|---|
| Up case | $aS_u + be^r = U$ |
| Middle case | $aS_m + be^r = M$ |
| Down case | $aS_d + be^r = D$ |

We wish to solve for $a$ and $b$. But in general that is impossible, since we have more equations to satisfy (three) than unknowns (two). To put it another way, we need a three-dimensional solution space, and what we have is two-dimensional.

However, all is not lost. Our observation suggests that we should add another financial instrument. Let us begin by adding a bond with a different interest rate, $R$. Thus, our new portfolio consists of

$a$      units of the stock $S$

$b$      units of bond 1 (price is \$1 at an interest rate $r$).

$c$      units of bond 2 (price is \$1 at an interest rate $R$).

Again, we set up the portfolio value at $t = 1$ in the three scenarios:

| | |
|---|---|
| Up case | $aS_u + be^r + ce^R = U$ |
| Middle case | $aS_m + be^r + ce^R = M$ |
| Down case | $aS_d + be^r + ce^R = D$ |

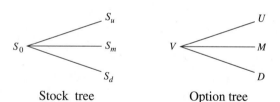

**FIGURE 2.12**

Trinomial model         Stock tree         Option tree

Now we have three equations and three unknowns. But look at the second and third terms on the left-hand side. They are identical in all three equations. So, again we cannot solve for $a$, $b$, and $c$ in general. Again, our solution space is only two-dimensional (or the space of the column vectors is two-dimensional), and we need three dimensions.

We still have one more strategy. Let us add another stock, $P$, which also takes one of exactly three values, $V_u$, $V_m$, and $V_d$, at $t = 1$. Our new portfolio consists of

$a$    units of the stock $S$

$b$    units of the bond (price is \$1 at an interest rate $r$)

$c$    units of stock $P$

We set up the portfolio value at $t = 1$ and we make the following

### Enormous Assumption

**1.** When $S \rightarrow S_u$, $P \rightarrow P_u$

**2.** When $S \rightarrow S_m$, $P \rightarrow P_m$

**3.** When $S \rightarrow S_d$, $P \rightarrow P_d$.

Thus, the portfolio values in the three cases become

$$aS_u + be^r + cP_u = U$$
$$aS_m + be^r + cP_m = M$$
$$aS_d + be^r + cP_d = D$$

We now have three independent equations and three unknowns, so we can solve for $a$, $b$, and $c$. Having done so, we can determine the initial portfolio value:

$$\Pi_0 = a \cdot S + b + cP$$

But look back at our Enormous Assumption. We have won a hollow victory. First, it does not seem reasonable that if we wish to price an option on Ford, we should be using an additional stock such as International Paper. Second, it is even more unreasonable to assume that Ford and International Paper move up, down, or sideways in tandem. We do not have to assume that $P_u$, $P_m$, and $P_d$ occur in any particular order, but the linked, coupled, or tandem movement is just too much to swallow.

## REVIEW EXERCISES

**1.** Find the derivative price given the following data:

|     | $S_0$ | $S_u$ | $S_d$ | U  | D   | r    | $\tau$ |
|-----|-------|-------|-------|----|-----|------|--------|
| (a) | 100   | 20    | 80    | 18 | -10 | 0.06 | 1 yr.  |
| (b) | 60    | 90    | 50    | 10 | 0   | 0.05 | 6 mo.  |
| (c) | 50    | 60    | 45    | 5  | 0   | 0.05 | 1 mo.  |
| (d) | 40    | 50    | 20    | 40 | -10 | 0.06 | 8 mo.  |
| (e) | 90    | 100   | 80    | 20 | -10 | 0.05 | 3 mo.  |
| (f) | 100   | 120   | 90    | 0  | 20  | 0.05 | 1 yr.  |

2. Assume that the derivative price in part (a) of Exercise 1 is quoted at $8. What should you do? Estimate (compute) your riskless profit.

3. Repeat Exercise 2 for part (b) of Exercise 1, where the derivative is offered at $3.

4. Repeat Exercise 2 for part (c) of Exercise 1, when the derivative is selling at $2.

5. Assume that you are the dealer and that a client wishes to purchase 1,000,000 options of the stock described by part (b) of Exercise 1 at a price of $0.10 over the *fair* market price. How do you hedge the position? Determine your net profit.

# CHAPTER 3

## TREE MODELS
## FOR STOCKS
## AND OPTIONS

*October is one of the peculiarly dangerous
months to speculate in stocks.*
*    The others are July, January, September, April,
November, May, March, June, December,
August, and February.*

Mark Twain

### 3.1   A STOCK MODEL

We used a binomial tree model for stock prices in the previous chapter to make
several calculations. We will extend these calculations to a more realistic multiperiod
model, while still maintaining simplicity.

$$S_0 = \text{stock price at time } t = 0$$

As before, our stock can move to only one of two prices over one unit of time. The
possibilities are indicated by the trees shown in Figures 3.1 and 3.2. Notice that the
future values are multiples of the present value. We assume that the **stock parameters,** $u$, $d$, and $p$, are given. Later, we will consider methods for determining these
values from real-world data.

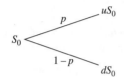

**FIGURE 3.1**
Binomial stock tree

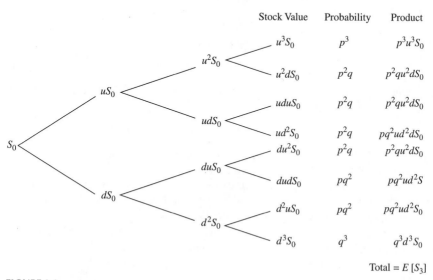

**FIGURE 3.2**
Stock tree in exploded form

The price, $S_1$, at time $t = 1$ is obtained by multiplying $S_0$ by either $u$ or $d$ and it is natural to use $u > 1$ and $d < 1$. We say that the *up move* occurs with probability $p$ and that a *down move* occurs with probability $q = 1 - p$. Before going further, let us calculate $E[S_1]$, the expected value of $S_1$.

Our choice of $S_1$ values include $S_0$ as a factor, so that we obtain

$$E[S_1] = p \cdot uS_0 + q \cdot dS_0$$
$$= (pu + qd)S_0$$

The term $pu + qd$ measures the *drift* of the stock price.

For $pu + qd > 1$ the price drifts up on balance

For $pu + qd < 1$ the price drifts down on balance

For $pu + qd = 1$ the price has no drift

Note that if $p = q = \frac{1}{2}$, then the drift is the arithmetic average, $(u + d)/2$, of $u$ and $d$. Let us consider this calculation when we are faced with several time periods for our stock.

### 3.1.1   Recombining Trees

We first write the tree in *exploded form*, as shown in Figure 3.2. We wish to find $E[S_3]$, and we have all the information at our fingertips:

$$E[S_3] = \text{sum of product column}$$
$$= [(pu)^3 + 3qd(pu)^2 + 3(qd)^2 pu + (qd)^3]S_0$$

after we collect terms. But we can write the last expression as

$$(pu + qd)^3 \cdot S_0$$

This last expression shows us clearly what is happening.

$$E[S_{k+1}] = (pu + qd)E[S_k] \tag{3.1}$$

In other words, each time period multiplies the expected value by the drift term, $pu + qd$. This implies that

$$E[S_k] = (pu + qd)^k S_0$$

Notice that the graph of the exploded stock tree grows at a stunning rate. The number of nodes after $k$ time periods is, in fact, $2^k$.

We will use a compressed version of this tree that is more tractable. Our exploded tree distinguished between the nodes $u^2 dS_0$ and $udu S_0$. It is quite reasonable to combine these nodes since they refer to the same stock value.

We arrive at the tree in Figure 3.3. Clearly this tree is more tractable. But can we use it to find $E[S_3]$, for example? Our ultimate goal is to use such trees to price options and derivatives and to describe the behavior of portfolios.

### 3.1.2   Chaining and Expected Values

We will start with $E[S_k]$ as an illustration of a general approach. The technique being introduced is called **chaining** or **backward induction.**

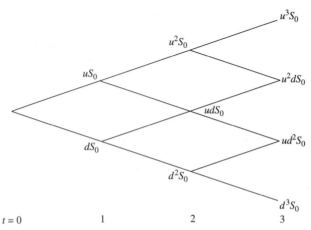

**FIGURE 3.3**
Compressed stock tree

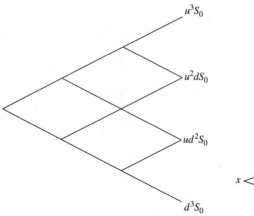

**FIGURE 3.4**
Stock price tree with last column completed

**FIGURE 3.5**
General stock price subtree

We begin with our tree stripped down to the column of final stock values, $t = 3$, as shown in Figure 3.4.

Our goal is to fill in the blank *nodes* with expected values, instead of the actual stock values. Suppose we had the truncated tree shown in Figure 3.5. If $x$ is the expected value of future values $a$ and $b$, then

$$x = p \cdot a + q \cdot b$$

and we call this the **chaining rule.** Let us apply this rule to fill column 3 in, corresponding to $t = 2$. The tree in Figure 3.6 results.

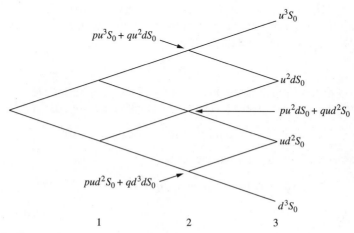

**FIGURE 3.6**
Stock price tree with two columns completed

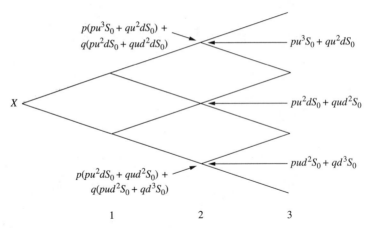

**FIGURE 3.7**
Stock price tree with three columns completed

### Important Remarks

1. Please note that we are not just repeating or undoing the steps used in Figure 3.3. The new nodes are different.
2. A clever, astute reader might, at this point, exclaim, "This can't work; you are ignoring the fact that the term $u^2dS_0$ appears three times in the exploded tree, Figure 3.2, and only once in Figure 3.6." If you did, this is a very good point and we commend you for your insight. If we proceed, this question will be answered.

We go on to column 2 ($t = 1$) and obtain the tree shown in Figure 3.7.

We have just one node to go; namely, to compute the value $X$. We claim that the node entry is precisely $E[S_3]$. This simple answer for $X$ is part of a pattern that emerges from the entries in Figure 3.7. As a first instance of the pattern, consider the topmost entry of the column for $t = 2$. The stock **value** at this node is $u^2S_0$. Suppose we factor out this quantity from the entry indicated by the arrow pointing to this node. We then can write the node entry as

$$u^2 S_0 (pu + qd)$$

As a second instance, look at the lowest entry in the column for $t = 1$. The stock value for this node is $dS_0$. If we factor out the stock value from the entry, we get

$$dS_0[p^2u^2 + 2pud + d^2q^2] = dS_0(pu + dq)^2.$$

Is the pattern clear? Each entry has the form

$$(\text{stock value at node}) \cdot (pu + dq)^{(\text{number of remaining columns})}$$

Thus

$$E[S_3] = S_0(pu + dq)^3$$

To verify the generality of the pattern, assume that it holds for the $k$th column of some stock tree. If we compute the entries for column $k - 1$ using the chaining rule, we find that the pattern is preserved.

## EXERCISES

**1.** Between one day and the next, the price of a hypothetical stock can increase 50% (with probability $\frac{1}{3}$) or decrease 10% (with probability $\frac{2}{3}$). If the stock is priced at $2.00 at the start of trading on Monday, what is the expected value of the stock at the start of trading on Thursday? Obtain your answer by filling in the nodes of a recombining tree.

**2.** Assume that a stock model has $u = 1.2$, $d = 0.8$, $p = 0.6$, and we know that $E[S_2] = 27.15$. Find $S_0$ and all possible values for $S_1$ and $S_2$.

**3.** If a stock model happens to have the parameters $u = 1.134$, $d = 0.8$, $p = 0.6$, and $S_2 = 6.38, 4.52$, or $3.2$, then when the nodes of the stock tree are filled in, we obtain the original stock prices. Explain why this is true.

## 3.2 PRICING A CALL OPTION WITH THE TREE MODEL

**Example 1** We assume

| | |
|---|---|
| $S_0 = 100$ | $u = 1.1$ |
| $X = 105$ | $d = 0.9$ |
| $r = 0.05$ | $p = 0.85$ |

The option expires at $t = 3$. There are three time periods for the stock. We first construct the price tree for our stock model (Figure 3.8).

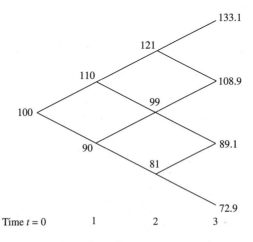

**FIGURE 3.8**
Stock price tree

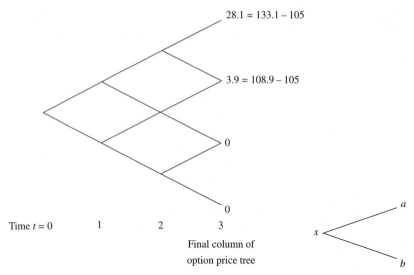

**FIGURE 3.9**
Option price tree with last column filled

**FIGURE 3.10**
General option subtree

We next draw the option price tree, but we put in only the values for $t = 3$ (Figure 3.9).

Next, use chaining to fill in the tree. Suppose we are faced with the subtree in Figure 3.10. We must first assign probabilities $q$ and $1 - q$, determined by arbitrage pricing, to our subtree. Recall that equation (2.5) gives the value

$$q = \frac{e^{r\tau}S_0 - S_d}{S_u - S_d}$$

Our imposed pattern, $S_u = uS_0$ and $S_d = dS_0$, allows us to use the *same q-value* at any part of our tree since

$$q = \frac{e^{r\tau} - d}{u - d} \tag{3.2}$$

Thus, in Example 1,

$$q = \frac{e^{0.05} - 0.9}{1.1 - 0.9} = 0.7564$$

Furthermore, the pricing equation (2.4) requires us to discount. Consequently, the chaining step becomes

$$x = e^{-r\tau}[\, qa + (1 - q)b \,] \tag{3.3}$$

Returning to the immediate example, this chaining step gives a top entry in the $t = 2$ column equal to

$$e^{-0.05}[(0.7564)28.1 + (0.2436)3.9] = 21.12$$

Filling in the rest of the tree, we arrive at Figure 3.11. So, the value of the option today is $11.87.

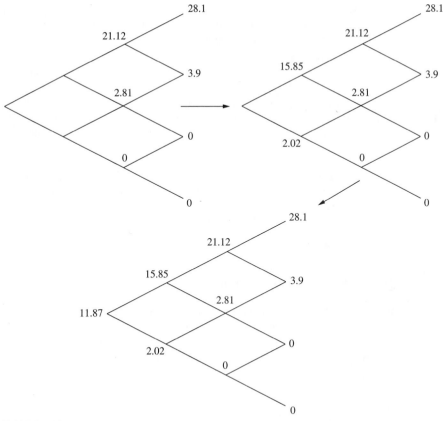

**FIGURE 3.11**
Completed option tree

Since this value is determined by the column 3 entries of our option tree, one might ask why we had to fill in all the nodes to get to the one node we are interested in. In fact, there is a shortcut. There is a one-step, expected value formula for the result, $11.87. This simple procedure is available because the value of this European option is nailed down once we know which final node we arrive at. However, we are about to see that for many options, the *route we take* to reach the final node affects its value. There is no simple, one-step procedure to obtain the price of these more complex options.

On the other hand, the **chaining method** is flexible enough to price American options, barrier options (knockout options) and interest rate options. These more complex options require us to look at path-dependent financial quantities.

## EXERCISES

**1.** Assume that a stock model has the parameters $u = 1.7$, $d = 0.8$, $S_0 = 120$. A European call option, expiring at $t = 3$, has an exercise price $X = 115$ and the interest rate is 0.06 over each period. Price this option at $t = 0$ using the chaining method.

2. Assume that a stock model has the same parameters as in Exercise 1. A European call option, expiring at $t = 3$, has an exercise price $X = 140$ with the same interest rate as before. Price this option at $t = 0$ using the chaining method.

3. A stock model has the parameters $u = 2.0$, $d = 0.5$, $S_0 = 16$. A European call option, expiring at $t = 4$, has an exercise price $X = 20$ and the interest rate is $0.1$ over each period. Price this option at $t = 0$ using the chaining method.

4. Verify that the following *one-step* method gives the correct answer for exercise 3: Draw the complete, 4-period stock tree. Each stock node, at the expiration date, will be used to compute a probability. To obtain this probability for the node $S_4 = 64$ we will multiply $4 \times q^3 \times (1 - q)$. The four is used because four paths arrive at this node. $q^3$ is used because any path has three up moves.

   Each of the five probabilities is found by multiplying the number of paths times $q$ to the number of up moves times $1 - q$ to the number of down moves. To complete the pricing of the call, take the call's profit, $S_4 - X$ or $0$, at each node, multiply it by the node probability, and sum the results. Finally, multiply this sum by $e^{-0.1 \times 4}$. This is the call price.

*Hint:* The number of paths to the nodes is respectively 1,4,6,4,1.

## 3.3 PRICING AN AMERICAN OPTION

A European option may be exercised only at the expiration date, but an American option can be exercised at any time prior to its expiration. How does "early exercise" affect the value of the option? In this section we will show how to price an American put option. We will use the same stock from the previous section. Here are the data again:

$$S_0 = 100 \qquad u = 1.1$$
$$X = 100 \qquad d = 0.9$$
$$r = 0.05 \qquad p = 0.85$$

Option expires at $t = 3$

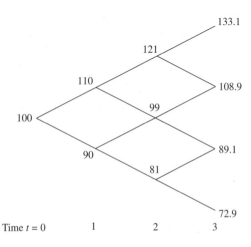

**FIGURE 3.12**
Stock price tree          Time $t = 0$          1          2          3

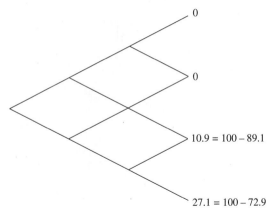

**FIGURE 3.13**
American put option value tree

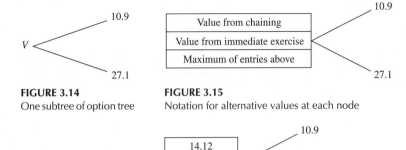

**FIGURE 3.14**
One subtree of option tree

**FIGURE 3.15**
Notation for alternative values at each node

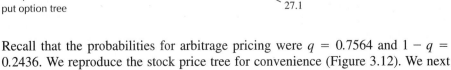

**FIGURE 3.16**
Completed node of American
put option tree

Recall that the probabilities for arbitrage pricing were $q = 0.7564$ and $1 - q = 0.2436$. We reproduce the stock price tree for convenience (Figure 3.12). We next fill in the put option value for just the last column (Figure 3.13).

Let us start with subtree shown in Figure 3.14. How should we value $V$? There are two choices. We could exercise the option at this point ($t = 2$) or hold it for one more time period (till $t = 3$). Our strategy will be to assign values to each alternative and then choose the maximum for $V$. Use the notation shown in Figure 3.15 to record the possible values and their maximum. The chaining value is

$$\text{Chaining value} = e^{-0.05}[\ 10.9(0.7564) + 27.1(0.2436)\ ] = 14.12$$

Our subtree entries are then as shown in Figure 3.16. Note that we do *not* discount the value, 19, because we receive the funds immediately. There is no waiting. The subsequent chaining calculations using this node are computed using the entry in the *"Maximum"* box. The tree, completed in the same manner, is shown in Figure 3.17. Consequently, our put option is worth $2.74 today. If it were not for the "early exercise" privilege, it would be worth considerably less.

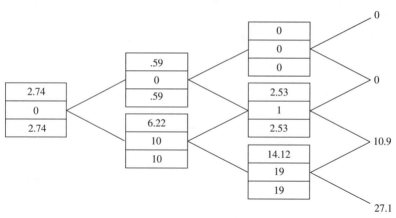

**FIGURE 3.17**
Completed tree for American put option

## EXERCISES

**1.** In this exercise you are given the number of periods, $n$; the initial stock price, $S_0$; the up and down multipliers, $u, d$; and the short-term interest rate, $r$. Use this information to find the price of a European call option at time 0.

|     | $n$ | $S_0$ | $u$ | $d$ | $X$ | $r$ |
|-----|-----|-------|-----|-----|-----|-----|
| (a) | 2 | 100 | 1.1 | 0.9 | 95 | 0.05 |
| (b) | 3 | 80 | 1.2 | 0.8 | 100 | 0.04 |
| (c) | 4 | 60 | 1.3 | 0.8 | 75 | 0.06 |
| (d) | 4 | 50 | 1.2 | 0.9 | n 45 | 0.03 |
| (e) | 4 | 40 | 1.1 | 0.7 | 40 | 0.05 |
| (f) | 5 | 110 | 1.4 | 0.7 | 120 | 0.06 |
| (g) | 5 | 90 | 1.3 | 0.9 | 80 | 0.04 |

**2.** Use the same values for $n$, $S_0$, $u, d$, $X$, and $r$ as in Exercise 1, but find the value of a European put associated with these parameters.

**3.** Use the same values for $n$, $S_0$, $u, d$, $X$, and $r$ as in Exercise 1, but find the value of an American put associated with these parameters.

**4.** Suppose that a put option that expires after one period is based on the stock model shown in Figure 3.18. Assume that $q = 0.8$ and $e^{rt} = 1.06$. Solve for exercise prices $X$, so that $90 < X < 110$ and an early exercise is more profitable than selling the put option.

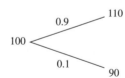

**FIGURE 3.18**
Stock model for Exercise 4

## 3.4 PRICING AN EXOTIC OPTION—KNOCKOUT OPTIONS

For this example, use the stock from the previous section:

| | |
|---|---|
| $S_0 = 100$ | $u = 1.1$ |
| $X = 105$ | $d = 0.9$ |
| $r = 0.05$ | $p = 0.85$ |
| $\tau = 3$ | Three time periods |

Our option will be a European call option expiring in three years with a strike of $105. This is similar to the previous examples, but there is one twist. This option is a **knockout** with the barrier set at $95. That is, if the stock price **ever** goes below $95, the option is worthless no matter what the stock price is at $t = 3$.

First, we present the stock price tree again (Figure 3.19), with the barrier delineated by the dashed line. Next, we put in the option price tree with the dashed line indicating those nodes above and below the barrier (Figure 3.20).

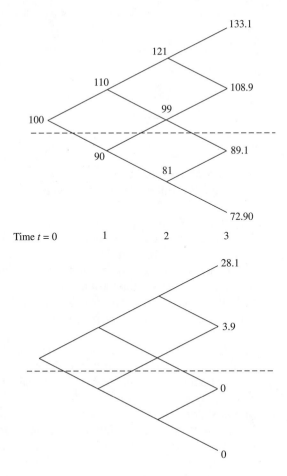

**FIGURE 3.19**
Stock price tree
with knockout barrier

**FIGURE 3.20**
Knockout option tree

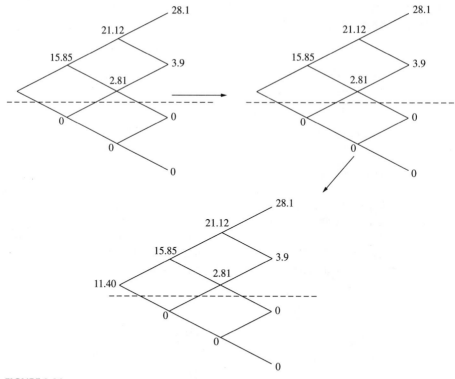

**FIGURE 3.21**
Sequence of node entries in knockout option tree

***Barrier Rule*** We compute the value of the option by chaining and discounting as usual. However, we assign the value 0 to any node lying below the barrier (dashed line).

In the chaining procedure, we use our arbitrage probabilities $q = 0.7564$ and $1 - q = 0.2436$. The values assigned to the nodes for $t = 2$ and $t = 3$ are the same as the nonbarrier case we calculated earlier. The barrier effect does not show up until the $t = 1$ node. The sequence of node entries is indicated in Figure 3.21.

We arrive at a price of $11.40 for the knockout option. Note that this option is less expensive than the "plain vanilla" option we evaluated in Section 3.1. That fits our intuition, since the option will pay off in cases where the knockout will not.

As a check on our method, we present a second method for calculating the value of the knockout option. The final value of this option depends on the **path** of the stock, not just on the final stock value. We describe the path by using $u$'s and $d$'s to indicate whether the path is going up or down. Read from left to right.

With this symbolism, $duu$ means the stock went down, then up, then up again. Paths whose final values are above the barrier are $uuu$, $uud$, and $udu$. Paths whose final values are below the barrier are $ddd$, $duu$, $dud$, $ddu$, and $udd$.

One path, *uuu*, leads to a payoff of $28.10 for the option, and two paths, *uud* and *udu*, lead to a payoff of $3.90 for the option. The reader is encouraged to trace out the various paths on the tree with colored markers. From our earlier discussion of the exploded tree it should be clear that

$$\text{Pr[Path } uuu \text{ occurs]} = q^3 = 0.7564^3 = 0.4328$$

and

$$\text{Pr[Path } uud \text{ or } udu \text{ occurs]} = 2q^2(1 - q) = 2(0.7564^2)(0.2436) = 0.2787$$

The expected value for the barrier option is obtained by summing over paths (and discounting over three years). The result is

$$e^{-0.05 \cdot 3} E[\text{ Paths }] = e^{-0.15}[(0.4328)28.1 + (0.2787)3.9]$$
$$= 11.40$$

This value confirms our earlier calculation.

**Remarks** The path computation was tractable given the small number of time steps and paths, but for $k$ time steps, the number of paths increases as $2^k$. Separating the "live" from the "dead" becomes onerous.

Barrier options cost less than the standard European option, as we have just seen. The price reduction is greatest when the barrier is set close to the prevailing price, because it becomes less likely that the instrument will survive the daily price swings.

The following table illustrates this assertion:

$$S_0 = 145, X = 145, r = 0.06, \text{dividend rate} = 3\%,$$
$$\text{volatility} = 29.5\%, \tau = 6 \text{ months}.$$

| Type of call option | Barrier | Price |
|---|---|---|
| Standard European | | 12.88 |
| Down and out | 130 | 10.48 |
| | 110 | 12.83 |
| Up and out | 160 | 0.17 |
| | 190 | 4.46 |

These figures were obtained from the CIBC (*Financial Products Class Notes* by Charles Smithson with William Chan).

The more common barrier options are those for which the same asset price serves as both the underlying price and the trigger (barrier) price. These are called "inside" barriers. Barrier options can also be created where the asset price used for the trigger is different from the underlying price; these are called "outside" barriers. For example, in the Japanese equity market, when the yen/U.S. dollar was trading at a ratio of 80.00, portfolio managers bought calls on the Nikkei 225 when it was trading in the 15,000 range. These calls knocked out if and only if the yen/U.S. dollar renewed its low, in this case in the 79.00 range.

Again, the embedding of the knockout feature significantly reduced the price of such calls. This structure also took full advantage of the Japanese equity market's sensitivity to the U.S. dollar. A stronger dollar (cheaper yen) is likely to fuel a recovery in the value of Japanese exporter stocks that would likely drive the overall market higher.

Barrier options are particularly attractive for hedge funds. Hedge funds are allowed to sell short and purchase options; they attempt to achieve the biggest bang for their buck. When a barrier is breached in their favor, the funds make a hefty return on the relatively small amount of money they pay for knockins or knockouts.

The problem comes when prices move close to the barrier. The buyer of a knockout option has a big incentive to keep prices below the barrier. The option's seller has an equally big incentive to keep prices above the barrier. In the case of a huge and liquid market, such as foreign exchange, a single investor might judge it too risky and impractical to try to move prices. But if enough big investors have the same interests, they can certainly exert significant pressure on the market.

Emerging markets that are small and illiquid are particularly vulnerable to this behavior. In late 1994 and early 1995, Merrill Lynch, an American investment bank, and a fund managed on behalf of Michael Steinhardt slugged it out in the market for Venezuelan bonds. The fund owned a knockin option, which becomes active if the price goes above the barrier, and was trying to push up prices by buying huge quantities of bonds. Merrill, which had sold the option, used all of its muscle to keep them below the barrier. Trading volumes in the otherwise obscure market soared. At the height of the battle, some $1.5 billion of the almost $7 billion of outstanding Venezuelan Brady bonds changed hands, pushing up prices by 10%. Merrill Lynch eventually won this titanic struggle, but at a considerable cost.

## EXERCISES

1. Use the stock tree in Figure 3.22, with a down-and-out barrier set at $65, to find the call option price at $t = 0$ with $X = \$50$ and $r = 0.06$.

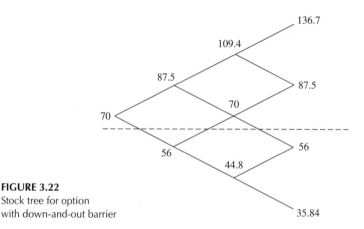

**FIGURE 3.22**
Stock tree for option
with down-and-out barrier

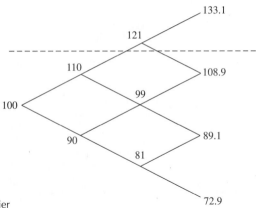

**FIGURE 3.23**
Stock tree for option
with up-and-out barrier

2. Find the down-and-out call price at $t = 0$ with the stock data in Exercise 1 if the barrier value is changed to \$55.
3. Use the stock tree in Figure 3.23, with an up-and-out barrier set at \$115, to find the call option price at $t = 0$ with $X = \$85$ and $r = 0.06$.

## 3.5  PRICING AN EXOTIC OPTION — LOOKBACK OPTIONS

Consider a **look back option** with a three-month expiration. At the end of three months, the buyer is paid the maximum value of the stock over the three-month period. The appropriate nature of the name "lookback" should be clear. Note that to determine the value of the option at expiration, you need to know more than the final stock value. You need to know where the stock has been at every point in time (except in a few extreme cases). To clarify the concept, we consider an example.

**Example**  Our stock has value 100 at $t = 0$. We are looking at a binomial model with $S_0 = 100$, $d = 0.9$, $u = 1.2$, and $r = 0.05$. We begin by drawing the stock price tree for the three-month period with time steps of one month, as shown in Figure 3.24.

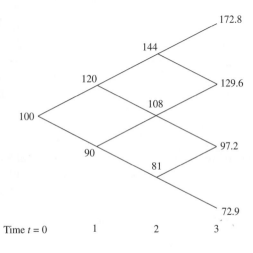

**FIGURE 3.24**
Stock tree for lookback
option example

Time $t = 0$      1      2      3

**TABLE 3.1**
**Paths and Their Maximum Values**

| Path | Maximum value |
|------|---------------|
| *uuu* | 172.8 |
| *uud* | 144 |
| *udu* | 129.6 |
| *duu* | 129.6 |
| *dud* | 108 |
| *ddu* | 100 |
| *udd* | 120 |
| *ddd* | 100 |

**TABLE 3.2**
**Probabilities of Paths**

| Path | Probability | Maximum value |
|------|-------------|---------------|
| *uuu* | $q^3$ | 172.8 |
| *uud* | $q^2(1-q)$ | 144 |
| *udu* | $q^2(1-q)$ | 129.6 |
| *duu* | $q^2(1-q)$ | 129.6 |
| *dud* | $q(1-q)^2$ | 108 |
| *ddu* | $q(1-q)^2$ | 100 |
| *udd* | $q(1-q)^2$ | 120 |
| *ddd* | $(1-q)^3$ | 100 |

To determine the price for the lookback, we must list all paths and the maximum value of the stock on each path. It is very helpful to draw all the paths in different colors. The paths are listed in Table 3.1.

Next, we need the pricing probability $q$ from equation (3.2):

$$q = \frac{e^{r\tau} - d}{u - d} = \frac{e^{0.05/12} - 0.9}{0.3}$$

$$= 0.34725$$

Next, we must compute the "probability" associated with each path. We list the results in Table 3.2. To find $E[V_3]$, the "expected value" of the option at time $t = 3$, we multiply each entry of the "Probability" column by the corresponding entry in the "Maximum value" column and then sum to obtain

$$E[V_3] = 115.314$$

Finally, we discount $E[V_3]$ to find the price of the option at time $t = 0$, since the pricing equation (2.6) includes discounting. Thus, we obtain

$$V_0 = 115.314 \cdot e^{(-0.05)3/12}$$

$$= \$113.88$$

---

The calculation of this value was a bit tedious. Because of the nature of the option, we must look at path behavior in order to assign an option value at the expiration date. You should verify that, if we were to draw the *exploded tree* for the option values, we could fill in the nodes according to our chaining rule and obtain the same answer.

The tabular approach used in the previous example, or an exploded tree leads to lengthy calculations as the number of periods increase. If there are $n$ time periods, there are $2^n$ paths. If we were working with a one-year lookback with monthly time intervals, then $n = 12$ (for the months), and hence there would be $2^{12} = 4096$ paths! Although a computer can handle such cases with ease, it would be helpful to have a less computational approach. In fact, the case of a one-year lookback using weekly periods ($n = 52$) would overwhelm even the fastest computers.

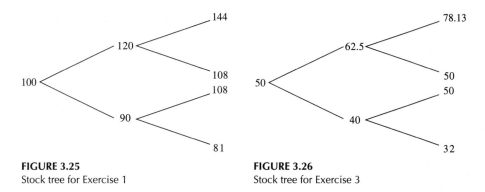

**FIGURE 3.25**
Stock tree for Exercise 1

**FIGURE 3.26**
Stock tree for Exercise 3

## EXERCISES

1. Find the price of a lookback option for the stock tree in Figure 3.25. You may use chaining.
2. Use the path probability method of this section to verify the price you obtained in Exercise 1.
3. Find the price of a lookback option for the stock tree in Figure 3.26.

## 3.6   ADJUSTING THE BINOMIAL TREE MODEL TO REAL-WORLD DATA

Mathematical models in finance (or in any area) are intended to help us understand the present and predict the future. If our models are to succeed they must harmonize with the real world.

Let us apply these comments to our binomial model for the stock prices, shown in Figure 3.27. How do we choose $p$, $u$, and $d$ ?

We wish to match these parameters to the important components in the stock's behavior. The two components that immediately leap to mind are drift and volatility. If we regard the stock price after a brief time period, $\Delta t$, as a random variable, $S$, then the *relative return* is the ratio

$$\frac{S - S_0}{S_0}$$

The *drift parameter*, $\mu$, measures the average percent change in the stock price over time. We hypothesize that

$$\mu \Delta t = E\left[ \frac{S}{S_0} - 1 \right] = E\left[ \frac{\Delta S}{S_0} \right] \tag{3.4}$$

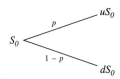

**FIGURE 3.27**
Binomial model of stock prices

That is, $\mu\Delta t$ is the average relative return. So $1 + \mu\Delta t$ is the mean of the stock ratio

$$\frac{S}{S_0}$$

At first glance, $1 + \mu\Delta t$ looks strange. But think of the formula for compound interest,

$$P = P_0(1 + r)$$

The term $1 + \mu\Delta t$ corresponds to the term $(1 + r)$, and $\mu\Delta t$ is analogous to $r$.

The *volatility parameter*, $\sigma$, measures the randomness of the relative return. We hypothesize that it determines the variance:

$$\sigma^2\Delta t = E\left[\left(\frac{S}{S_0} - 1 - \mu\Delta t\right)^2\right] \tag{3.5}$$

If we had a financial instrument, $P$, with no volatility (a money market fund), then

$$dP = \mu P \cdot dt$$

or, in differential form, $dP/P = \mu\, dt$. This matches the situation in equation (3.4). The term $S - S_0$ plays the role of $dP$. We see that

$$P(t) = P_0 e^{\mu t}$$

The behavior of $P$ is deterministic, and the graph of $P$ is the familiar exponential curve.

For the Bernoulli random variable $X$, summarized by Table 3.3,

$$\text{The mean } = E[X] = pa + (1 - p)b \tag{3.6}$$

and

$$\text{The variance } = E[(X - \mu)^2] = p(1 - p)(a - b)^2 \tag{3.7}$$

Let us apply this information to our stock tree, Figure 3.27. Then, according to our hypotheses,

$$1 + \mu\Delta t = E[S/S_0] = pu + (1 - p)d$$

and

$$\sigma\sqrt{\Delta t} = \sqrt{E[(S/S_0) - 1 - \mu\Delta t)^2]} = \sqrt{p(1 - p)}(u - d)$$

**TABLE 3.3**
**Bernoulli Random Variable**

| $X$ Value | Probability |
|-----------|-------------|
| $a$ | $p$ |
| $b$ | $1 - p$ |

*Implementation* To build a model we need two things:

**1.** A way to derive $u$, $d$, $p$ from the real-world values for $\mu$, $\sigma$
**2.** A way to determine $\mu$, $\sigma$

Let us attack them in that order.

There are several "models" for determining $u$ and $d$. We discuss a simple method that is frequently used for modeling stocks.

*Hull-White Algorithm* We set $p = \frac{1}{2}$ and determine $u$ and $d$ from the equations

$$\text{(i)} \quad \frac{u + d}{2} = 1 + \mu \Delta t$$

$$\text{(ii)} \quad u - d = 2\sigma \sqrt{\Delta t}$$

Note that equation (i) is just (3.6) with $p = \frac{1}{2}$, and equation (ii) is just (3.7). To summarize: Equations (i) and (ii) tell us how to define $u$ and $d$ given $\mu$, $\sigma$, and $\Delta t$.

The second step is just standard statistics. We assume

$$\begin{aligned} S_1 &= X_1 S_0 \\ S_{k+1} &= X_{k+1} S_k \end{aligned} \tag{3.8}$$

where the $X_k$ are independent Bernoulli random variables with $\Pr[\, S_k/S_{k-1} = u\,] = \Pr[\, S_k/S_{k-1} = d\,] = \frac{1}{2}$. This last sentence is just a wordy description of the tree in Figure 3.28.

From equation (3.7) we see that the reasonable estimators of $\mu \Delta t$ and $\sigma^2 \Delta t$ are

$$\bar{U} = \frac{1}{n} \sum_{k=1}^{n} (X_k - 1) = \frac{1}{n} \sum_{k=1}^{n} (S_k/S_{k-1} - 1)$$

and $s^2$, where

$$s^2 = \frac{1}{n-1} \left[ \sum_{k=1}^{n} (S_k/S_{k-1} - 1)^2 - n\bar{U}^2 \right].$$

The numbers $\bar{U}$ and $s^2$ are the sample mean and sample variance that are determined from the real-world data, $S_0, S_1, S_2, \ldots, S_n$. We obtain estimates for $\mu$ and $\sigma$:

$$\mu \approx \frac{\bar{U}}{\Delta t}$$

$$\sigma \approx \frac{s}{\sqrt{\Delta t}}$$

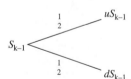

**FIGURE 3.28**
Stock model

**FIGURE 3.29**
Solution to Hull-White model

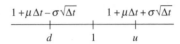

Once the data are used to estimate these drift and volatility parameters, we might decide to use a different value of $\Delta t$ for the time period of our binomial tree. We then solve for $u$, $d$ by replacing $\mu$ and $\sigma$ in equations (i) and (ii) with the estimates above. We obtain the following result, shown graphically as Figure 3.29.

$$u = 1 + \mu\Delta t + \sigma \sqrt{\Delta t}$$
$$d = 1 + \mu\Delta t - \sigma \sqrt{\Delta t}$$

In the Hull-White model, the $u$ and $d$ are symmetric with respect to the drift, $\mu$.

---

**Example**  To illustrate the material just discussed we present a toy model for Netscape® stock. The data were taken from the period May 4—May 18, 1998. Using a spreadsheet program, we enter a column of stock prices and compute the $X_j$'s and then $\bar{U}$ and $s$ with aid of some spreadsheet formulas (Table 3.4).

We can now take $\bar{U} = 0.992 - 1 = -0.008$ and $s = 0.033$ and determine $u$ and $d$. Our choice of $\Delta t$ for this tree example equals the $\Delta t$ for our data set. The formulas for $u$ and $d$ simplify to

$$u = 1 + \bar{U} + s$$
$$d = 1 + \bar{U} - s$$
$$u = 0.992 + 0.033 = 1.025$$
$$d = 0.992 - 0.033 = 0.959$$
$$S_0 = 27.25 \quad \text{(the price on May 18)}$$

The completed tree appears as Figure 3.30.

**TABLE 3.4**
**Netscape Stock, May 1998**

| Dates | Price | Ratio | Descriptive statistics | |
|-------|-------|-------|------------------------|---|
| 4  | 29.56  | 1.01911  | | |
| 5  | 30.125 | 0.94804  | Mean | 0.992383 |
| 6  | 28.56  | 0.96288  | Standard error | 0.010430 |
| 7  | 27.5   | 1.022727 | Median | 0.980898 |
| 8  | 28.125 | 0.975644 | Mode | |
| 11 | 27.44  | 0.98615  | Standard deviation | 0.032982 |
| 12 | 27.06  | 1.00240  | Sample variance | 0.001087 |
| 13 | 27.125 | 1.057695 | Kurtosis | 0.173060 |
| 14 | 28.69  | 0.973858 | Skewness | 0.764779 |
| 15 | 27.94  | 0.975304 | Range | 0.109646 |
| 18 | 27.25  |          | Minimum | 0.948049 |
|    |        |          | Maximum | 1.057695 |
|    |        |          | Sum | 9.923832 |
|    |        |          | Count | 10 |

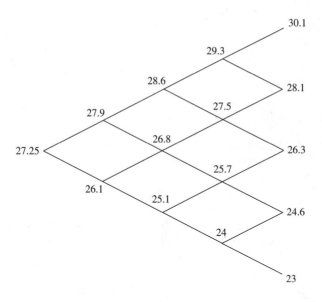

**FIGURE 3.30**
Price tree for Netscape: one-day time increments, $u = 1.025$,
$d = 0.959$

We could use larger or smaller time increments if we wished. For example, let us set $\Delta t = 7$, so we are looking at a week at a time. Then

$$u = 1 - 0.008(7) + 0.033\sqrt{7} = 1.031$$
$$d = 1 - 0.008(7) - 0.033\sqrt{7} = 0.8567$$

The completed tree is shown in Figure 3.31. Note that the prices "diverge" much faster with the weekly time steps.

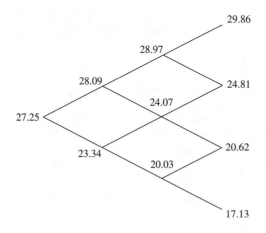

**FIGURE 3.31**
Netscape price tree: weekly time increments, $u = 1.031, d = 0.857$

*Remark* We had only 11 days of data for Netscape. In practice, we would want many more. A prudent forecaster would want to have many more data points in the past before attempting to estimate future price behavior.

## 3.7  HEDGING AND PRICING THE *N*-PERIOD BINOMIAL MODEL

The methods of this chapter suggest that the price of an option, based on an *N*-period binomial stock model, is completely determined by the option values at the final stage and the stock tree. We will illustrate why *any price* other than the tree-computed value allows a risk-free investment profit via arbitrage.

In other words, the computed price is the only one consistent with the lack of arbitrage opportunities. In this respect, the computed prices reflect real market conditions. Our illustrations will involve specific examples, and these provide a review of the methods we have seen in this chapter.

Let us verify the effectiveness of the delta hedge in the one-stage binomial process. We sell one call (or derivative) and buy *a* shares of stock where

$$a = \frac{U - D}{S_u - S_d}$$

Does this hedge really protect us against the uncertainty of the market?

Our initial cost is $aS_0 - C$, where $C$ is the price of the call. Note that

$$C = aS_0 + (U - aS_u)e^{-r\tau}$$

The quantity $aS_0 - C$ can be either positive or negative. In either case, we invest or borrow it at the risk-free rate, $r$.

Up state:  Value of position $= aS_u - U + (aS_0 - C)e^{r\tau}$
$$= aS_u - U + [aS_0 - (aS_0 + (U - aS_u)e^{-r\tau})]e^{r\tau}$$
$$= aS_u - U - (U - aS_u) = 0$$

Thus, we end up "flat," or even, in the up state, as we wished.

Down state:    Value of position $= aS_d - D + (aS_0 - C)e^{r\tau}$

But $aS_d - D = aS_u - U$, since that is how $a$ was defined. So, again, we see that we are "flat." In other words, we are completely hedged.

One might wonder how one could profit from a hedged position. In effect, a dealer under these circumstances would charge a small commission for a riskless profit.

We illustrate this hedging idea in an example that has several time periods.

---

**Example** We are given the stock price tree in Figure 3.32. We make no assumptions as to the origin of the tree. It could have been generated from an up-down price model, or it could have simply appeared out of thin air. We are also given the values for an

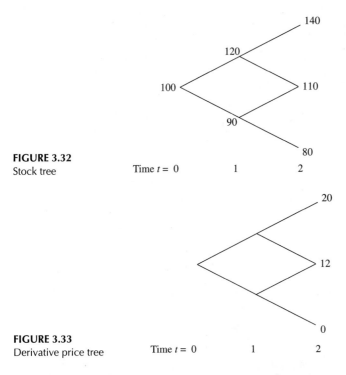

**FIGURE 3.32**
Stock tree

**FIGURE 3.33**
Derivative price tree

option or derivative at $t = 2$, displayed in Figure 3.33. In addition, we are given the risk-free rate. In this example, we suppose that $r = 0.04879$ so that $e^r = 1.05$.

***Step 1*** Using chaining, we fill in the derivative price tree. We have three methods for computing prices in this environment. We will not use $q$ values to fill in the derivative tree, because the $q$ value approach is convenient only when the stock tree is determined by $u$ and $d$ parameters. This tree is more general.

Instead, we use the game theory method of Chapter 2. Equation (2.1) states that

$$V = e^{-r\Delta t}(U - aS_u) + aS_0$$

where $a = (U - D)/(S_u - S_d)$. This calculation applies to any binomial subtree having the form of Figure 3.34.

Let us compute the $V$ value for each node of the derivative tree and record the $V$ value above the stock value at each node, as shown in Figure 3.35.

***Step 2*** Assume we have sold **one** derivative instrument (DI), based on the payoff at time 2, for $10.00756. We wish to hedge our sale to protect ourselves. Depending on the outcome, we face a payout of $12 or $20. We use the replicating-portfolio technique

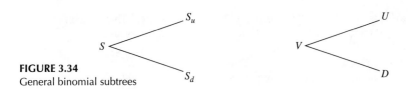

**FIGURE 3.34**
General binomial subtrees

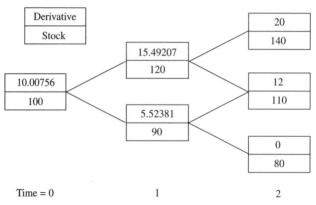

Time = 0                              1                              2

**FIGURE 3.35**
Combined stock and derivative price tree

(Section 2.3), to compute the required number of stock shares to hold at each time period.

Section 2.3 gives the answer for how many shares to hold at each node of our combined tree. Equation (2.8) is our familiar ratio,

$$a = \frac{\Delta V}{\Delta S} = \frac{U - D}{S_u - S_d}$$

So at time $t = 0$ we should purchase

$$\frac{\Delta V}{\Delta S} = \frac{15.492072 - 5.52381}{120 - 90} = 0.3322754$$

shares of stock. At time $t = 0$ we will also *borrow* some money as part of our hedge! We received \$10.00756 for the sale of the DI. However, in order to purchase the stock we will have to borrow

$$0.3322754 \times 100 - 10.00756 = \$23.21998$$

Our complete position at time $t = 0$ is summarized as

| Stock Owned | Owed to Bank |
|:---:|:---:|
| 0.3322754 | \$23.21998 |

At $t = 1$, we move to either the \$120 or the \$90 stock state, and at that point we must rehedge.

Assume that we move to the \$120 state. To rehedge our position, we calculate $\Delta V / \Delta S$.

$$\frac{\Delta V}{\Delta S} = \frac{20 - 12}{140 - 110} = 0.266666$$

Thus, we must sell $0.3322756 - 0.266666 = 0.06561$ shares. This *reduces* our debt position by $0.06561 \times \$120 = 7.8732$, and our new position consists of

| Stock Owned | Owed to Bank |
|:---:|:---:|
| 0.266666 | \$23.21998 \times 1.05 - 7.8732 = \$16.507779 |

Note that we have added the interest to the first year debt and paid off part of the loan with the sale of stock.

We now move to $t = 2$ and check to see whether this hedging strategy actually works. We assume that we end up in the $110 state. We sell the stock and pay off the DI holder and the bank. Let us set up a table summarizing these transactions:

| Positive (Asset) | Negative (Liabilities) |
|---|---|
| Stock: $0.266666 \times 110$ | DI: $12 |
| $= 29.33326$ | |
| | Bank: $16.507779 \times 1.05$ |
| | $= \$17.33316795$ |
| Total   29.33326 | 29.33316 |

Thus, the hedge is exact to roundoff error. We are completely and perfectly hedged. There is no risk in the DI sale. Check that the hedge protects the investor in the $140 state.

*Arbitrage* Note that if the derivative were priced at $11, we could make a riskless profit by selling the derivative and hedging the sale through stock purchases. On the other hand, if the derivative were priced at $9, we could achieve a riskless profit by buying the derivative and hedging the position through short selling the stock. Such transactions will drive the derivative price to $10.00756.

---

In summary, to determine the hedging position, we proceed as follows:

1. Find the stock price tree.
2. Calculate the derivative value tree through chaining.
3. Determine the stock hedge by starting at $t = 0$. The number of shares needed is $(U - D)/(S_u - S_d)$.
4. As you move on to $t = 1$ and the later periods, you must recalculate the number of shares of stock needed using Step 3.

This method works for the $N$-period binomial model.

*Implications* Does our ability to hedge precisely the $N$-period binomial model imply that we can hedge arbitrary derivatives (say, the continuous case)? No. Even though the binomial model offers a reasonable approximation to stock prices for large values of $n$, the rigid structure imposed on movements of the stock price does not allow us to make error-free hedge adjustments.

For example, when a stock price falls precipitously, its trading volatility tends to rise. To compensate for this while hedging, one should reduce the period between adjustments. Moreover, stock prices can and do gap up or down. For example, Microsoft was selling at $106 \frac{1}{4}$ on March 31, 2000, when news of the government's decision to proceed with antitrust prosecution was released. The stock immediately fell to $90 \frac{13}{16}$ on the next trading day, April, 3, 2000. It is not possible to hedge such moves. The binomial model is an imperfect model for most stocks, and hence a hedge based on this model will be imperfect.

## EXERCISES

1. In this exercise you are given the number of periods, $n$; the initial stock price, $S_0$; the up and down multipliers, $u, d$; and the short-term interest rate, $r$. Find the $\Delta$ hedging values for one European call option.

   This determination requires the selection of a path through the tree. For simplicity, we suggest considering each case, such as $udud$, but you may wish to explore other choices for path selections.

2. Use the same values for $n$, $S_0$, $u$, $d$, $X$, and $r$ as in Exercise 1, but find the $\Delta$ hedging values for one European put option.

|     | $n$ | $S_0$ | $u$ | $d$ | $X$ | $r$ |
|-----|-----|-------|-----|-----|-----|-----|
| (a) | 2   | 100   | 1.1 | 0.9 | 95  | 0.05 |
| (b) | 3   | 80    | 1.2 | 0.8 | 100 | 0.04 |
| (c) | 4   | 60    | 1.3 | 0.8 | 75  | 0.06 |
| (d) | 4   | 50    | 1.2 | 0.9 | 45  | 0.03 |
| (e) | 4   | 40    | 1.1 | 0.7 | 40  | 0.05 |
| (f) | 5   | 110   | 1.4 | 0.7 | 120 | 0.06 |
| (g) | 5   | 90    | 1.3 | 0.9 | 80  | 0.04 |

3. Use the values for $n$, $S_0$, $u$, $d$, $X$, and $r$ of parts (b) and (c) only. Find the $\Delta$ hedging values for an *American put* associated with these parameters.

# CHAPTER
# 4

## USING SPREADSHEETS TO COMPUTE STOCK AND OPTION TREES

*The future is already here. It's just unevenly distributed.*

William Gibson

## 4.1 SOME SPREADSHEET BASICS

In Chapter 3 we learned that each node in a stock tree (such as that shown in Figure 4.1) has a value, $S_k$, that is obtained by multiplying a neighboring value by one of two constants, $u$ or $d$.

The actual process of filling in these values is tedious. A spreadsheet is a very useful tool for eliminating most of these steps. To see how this works, let us look at a typical blank spreadsheet page that appears on the computer screen when a spreadsheet program is opened (Figure 4.2). Each "cell" in this grid can hold either a title such as "Stock Price," a numerical value, or a formula.

As a first step, enter the value 100 in cell J1. This is done by selecting the cell with the mouse, typing **100**, and hitting the Enter (or Return) key. As you type, you should see 100 first appear on a special line above the cells. After you hit Enter, the number will appear in cell J1.

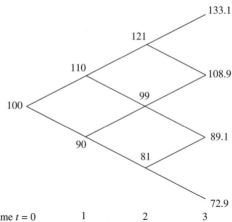

**FIGURE 4.1**
Stock tree

| | A | B | C | D | E | F | G | H | I | J |
|---|---|---|---|---|---|---|---|---|---|---|
| 1 | | | | | | | | | | |
| 2 | | | | | | | | | | |
| 3 | | | | | | | | | | |
| 4 | | | | | | | | | | |
| 5 | | | | | | | | | | |
| 6 | | | | | | | | | | |

**FIGURE 4.2**
Blank spreadsheet

Next, make a stock tree that branches down from J1. The next two nodes of the stock tree will be I2 and K2. To produce the entry for I2, select it with the mouse and type

**=.9*J1**

Although this **formula** appears on the special top line as you type, after you hit Enter the cell will show a numerical value, 90. The spreadsheet has retained this formula in cell I2.

This is a good point at which to pause. **Save** the spreadsheet you are creating, and give it a file name such as "Stock Tree" or "stock."

The formula we created involved the number $d = 0.9$. There is a simple way to make our stock tree more flexible so that it accepts other values of $d$. Let us agree that we will always enter the correct $d$ value in cell A7. In fact, enter **.9** in A7 and enter the title **d value** in cell B7 to remind us where our $d$ parameter is located. We wish to treat the initial stock value in the same manner. Enter 100 in **A1** and the title **Stock Price** in cell B1. To make the full titles visible, you should change the width of column **B**.

| | A | B | C | D | E | F | G | H | I | J | K |
|---|---|---|---|---|---|---|---|---|---|---|---|
| 1 | 100 | Stock Price | | | | | | | | 100 | |
| 2 | | | | | | | | | 90 | | |
| 3 | | | | | | | | | | | |
| 4 | | | | | | | | | | | |
| 5 | | | | | | | | | | | |
| 6 | | | | | | | | | | | |
| 7 | .9 | d value | | | | | | | | | |

**FIGURE 4.3**
Spreadsheet with six entries

To use the entries in column A, just replace the entry in cell J1 by the simple formula, **=A1**. Also, replace the formula in cell I2 with **=J1\*$A$7**. The numerical values will remain the same. The new feature is that if we change either the stock price or the d value, then the I and J columns change accordingly. Now your spreadsheet should look like Figure 4.3.

Why did we put dollar signs in the I2 formula? They signal the spreadsheet not to change the A7 address as we copy and paste this formula. The next step is to copy the formula by selecting the cell I2 and choosing a copy command. Then select the new cell, H3, and paste the formula there. Notice that the value there is 81. Continue to paste the *original formula* in cells G4, F5, and E6. These are locations down one row and left one column from the previous cell. If you check the formula entry for any of these cells, you will see that it multiplies the cell value of its upper right neighbor by the d value. So we have one edge of our stock tree filled, and this edge can accept any change of either the d value or the initial stock price.

We can fill in more of the stock tree if we use a u value. Let us agree that the u value will be entered into A6. In fact, enter the title **u value** in B6. It is good practice to **save** the spreadsheet file again since we have invested some effort into creating these cells and formulas. We suggest even saving a copy of this file with another file name, such as "Template". We will use these formulas again soon.

Now enter the u value, **1.1**, in A6. Next, select the cell that is down and to the *right* of the beginning node of our stock tree. This is K2. Type the formula **=J1\*$A$6** and hit Enter. Your spreadsheet should look like Figure 4.4.

You will enjoy completing the final step. Select the formula we just created in K2 and choose a copy command. Next, paste the formula into any cell that is down and to the right of a location containing a numerical value. A good place to start is at J3, and you can immediately paste at the locations K4 and L5 to see the sequence 99, 109, and 120 appear. Continue to fill out the entire stock tree this way. When partially filled out, the tree should appear like Figure 4.5.

If you succeeded in producing these values, save this sheet once more (still as "Stock tree" or "Stock") and save another copy under a different file name as well. This sheet will be valuable for working exercises that include finding the prices of options, too.

| | A | B | C | D | E | F | G | H | I | J | K |
|---|---|---|---|---|---|---|---|---|---|---|---|
| 1 | 100 | Stock Price | | | | | | | | 100 | |
| 2 | | | | | | | | | 90 | | 110 |
| 3 | | | | | | | | 81 | | | |
| 4 | | | | | | | 72.9 | | | | |
| 5 | | | | | | 65.6 | | | | | |
| 6 | 1.1 | u value | | | 59 | | | | | | |
| 7 | .9 | d value | | | | | | | | | |

**FIGURE 4.4**
Spreadsheet after *u* branch is begun

| | A | B | C | D | E | F | G | H | I | J | K |
|---|---|---|---|---|---|---|---|---|---|---|---|
| 1 | 100 | Stock Price | | | | | | | | 100 | |
| 2 | | | | | | | | | 90 | | 110 |
| 3 | | | | | | | | 81 | | 99 | |
| 4 | | | | | | | 72.9 | | 89.1 | | 109 |
| 5 | | | | | | 65.6 | | 80.2 | | 98 | |
| 6 | 1.1 | u value | | | 59 | | 72.2 | | 88.2 | | |
| 7 | .9 | d value | | | | | | | | | |

**FIGURE 4.5**
Spreadsheet with tree partially filled by pasting

## EXERCISES

**1.** Figure 4.1 has a stock tree with $u = 1.1$ and $d = 0.9$. Reproduce this stock tree with your spreadsheet and print the result.

**2.** Exercise 1 of Section 3.4 has a stock tree with $S_0 = 70$ and three time periods. Reproduce this stock tree with your spreadsheet and print the result.

**3.** Suppose that you edit your "Stock" spreadsheet in the following way. Instead of entering the *d* value into A7, you enter the formula $= 1/A6$ in A7. Now choose $S_0 = 100, u = 1.2$ in this new spreadsheet and print the result of six time periods. Explain why several stock values are repeated.

**4.** Use the edited spreadsheet of Exercise 3 to print the stock tree of Exercise 1 of Section 3.4. Explain why you cannot reproduce Figure 4.1 with this method.

## 4.2 COMPUTING EUROPEAN OPTION TREES

Recall from Section 3.2 that we can use a stock tree to immediately write out an option price, such as a call, on its date of expiration. The spreadsheet we created in Section 4.1 will be updated to perform the same calculation for us. It will have the dynamic feature that as soon as we change the *u* or *d* value, the option price will change automatically. Of course, our ultimate goal is to fill in the entire option tree to obtain earlier prices.

As a first step, let us decide where to locate our option tree on our spreadsheet page. We want to use the *same* page we created for Figure 4.2 in order to use the stock values from that tree. We will locate the option tree between rows 11 and 20. Any option entry will be in the same column as the stock price it is based on. This means that the top option entry will be in J11.

The possible call prices on the date of expiration depend on the number of steps in the stock tree. Record the number of steps in A3 and enter the title **# Periods** in B3. Also, we need a location for the *strike price*, so enter the value **105** in A2 and the title **Strike Price** in B2.

The stock values after five periods are on line 6 of the spreadsheet. So we will fill in the option values on line 16 (putting the top of our tree at J11). What are these option values? They are 0 if the stock price is lower than the strike price, and they are the difference between the price and the strike if the stock is higher than the strike. The spreadsheet has a formula for this.

The formula for K16 is

$$=\text{MAX(K6-\$A\$2,0)}$$

Enter this formula into K16 (you do not need to capitalize letters). You should **save** this spreadsheet and give it a different file name such as "Euro Call." The formula you just entered refers to K6 to obtain the stock value, and subtracts the number in A2 from it. If the difference is negative, it records a 0 in the cell. The built in MAX function is extremely convenient and helpful.

Next, copy this formula and paste it in several locations on line 16, in columns E, G, I, M, and O. Line 16 of your spreadsheet should look as shown in Figure 4.6.

We would like to use the chaining method, introduced in Section 3.2, to fill in the top portion of the option tree. This procedure relies on two quantities that were calculated in Section 3.2. Soon we will devise formulas for the correct *arbitrage pricing probability*, $q$, and the discount factor, but for now we will simply enter these values as 0.7564 in cell A12 and 0.95 in cell A11.

|    | E | F | G | H | I | J | K | L | M | N | O |
|----|---|---|---|---|---|---|---|---|---|---|---|
| 11 |   |   |   |   |   |   |   |   |   |   |   |
| 12 |   |   |   |   |   |   |   |   |   |   |   |
| 13 |   |   |   |   |   |   |   |   |   |   |   |
| 14 |   |   |   |   |   |   |   |   |   |   |   |
| 15 |   |   |   |   |   |   |   |   |   |   |   |
| 16 | 0 |   | 0 |   | 0 |   | 2.81 |   | 26.8 |   | 56.1 |
| 17 |   |   |   |   |   |   |   |   |   |   |   |

**FIGURE 4.6**
One row of option tree

You should also enter the titles **q** and **exp −r** in the adjoining cells, B12 and B11. For convenience, enter $1 - q$ in A13 by entering the formula **=1-$A$12** in that cell.

Now we can fill in J15 with the chaining value for the option. If you look at formula (2.6) in Section 2.3.3, you will see that we should average the cell entries in I16 and K16 and then discount this average. The formula we enter (in J15) is

$$\textbf{=\$A\$11*(\$A\$13*I16+\$A\$12*K16)}$$

This formula is really

$$e^{-r}[\,(1 - q)\text{I16} + q\text{K16}]$$

in disguise. The number 2.02 should appear in the cell.

We have to type this formula in *only once*. If we had to type it for every entry, we would probably reject spreadsheets as inefficient and unwieldy. We can copy this formula once and paste it at all the nodes of our option tree. If you do this, your topmost entry will be 19.7. This is the chaining value for the present price of the option.

***The Refined Spreadsheet***   Although our option-stock tree spreadsheet is certainly worth **saving**, it suffers from the defect that the discount factor and the $q$ value must be entered. Let us correct this defect. We will use A5 as a location to enter the interest rate; enter the corresponding title in B5.

To use this cell in our calculations, enter the title **exp r** in B10 and enter the following formula in A10:

$$\textbf{= EXP(\$A\$5)}$$

Now we will put the correct discount factor in A11 by replacing its entry with this formula:

$$\textbf{= 1 / \$A\$10}$$

Our final step is to put the formula for the arbitrage pricing probability, $q$, in A12. This probability is given in Chapter 2; see equation (2.5). The formula is

$$\textbf{= (\$A\$10-\$A\$7) / (\$A\$6-\$A\$7)}$$

Once you have entered this formula, any changes in the parameters $u, d, r$, the stock price, or the strike price will produce the correct tree. Think about this last statement. Your spreadsheet, once constructed, will carry out tedious computation quickly and effortlessly.

## EXERCISES

1. Use a "Euro Call" spreadsheet to find the call price given that the stock price at $t = 0$ is $120, the exercise price is $130, expiration is 5 periods into the future, the interest rate per period is 4.5%, $u = 1.05$, and $d = 0.88$. Record the option's price at the bottom of your spreadsheet. Then change the exercise price to $125 and to $120, and record the new option prices on your spreadsheet. Print your spreadsheet.

2. Use a "Euro Call" spreadsheet to find the call price given that the stock price at $t = 0$ is \$65, the exercise price is \$70, expiration is 5 periods into the future, the interest rate per period is 5%, $u = 1.1$, and $d = 0.91$. Record the option's price at the bottom of your spreadsheet. Then change the interest rate to 6% and to 7%, and record the new option prices on your spreadsheet. Print your spreadsheet. Why does the price increase as $r$ increases?

3. Use the "Euro Call" spreadsheet of Exercise 1 to find the call price given the same stock parameters and $X = 130$. However, the expiration is **8** periods into the future.

*Hint:* You must copy and paste lines 17 and 27 to produce 3 new time periods.

4. Use the "Euro Call" spreadsheet of Exercise 2 to find the call price given the same stock parameters and $X = 70$. However, the expiration is **7** periods into the future.

*Hint:* You must copy and paste lines 17 and 27 to produce new time periods.

## 4.3   COMPUTING AMERICAN OPTION TREES

Sometimes it is profitable to exercise an option before its expiration date. As stated in Section 1.2.2, an option that allows this possibility is called an *American option*. The option's price can be computed with the tree method, but, as was explained in Section 3.3, the node entries must be tested in a certain order so that the replicating-portfolio method applies to the tree.

We have an option tree spreadsheet, created in Section 4.2, that performs the chaining steps for a European call. It is easy to adjust its formulas to check for early exercise, but no new price emerges from this adjustment. There is a financial reason why early exercise for calls is unprofitable.[1] However, American puts are more interesting.

We will modify the option tree from Section 4.2 so that it prices **American puts**. Use line 16 of the spreadsheet to record the possible expiration prices for a put that expires after five periods.

What are these option values? They are 0 if the stock price is higher than the strike price, and they are the difference of the strike and the stock price if the stock is lower than the strike. To state this briefly, we may say that the value is 0 unless "strike" minus "stock" is positive and then we take the latter amount. The appropriate spreadsheet formula that expresses this rule, for cell K16, is

$$= \text{MAX}(\$A\$2\text{-}K6\,,\,0)$$

Recall that A2 is involved because that is where the strike price is stored. Enter this formula and **save** this spreadsheet, giving it a different file name such as "Amer Put." The formula you just entered refers to A2 to obtain the strike and subtracts the number in K6 from it. If the difference is negative, it records a 0 in the cell.

---

[1]Hull, J. C., *Options, Futures, and Other Derivatives,* 3d ed., Prentice Hall, Upper Saddle River, NJ. 1997, pp. 162–165.

| | E | F | G | H | I | J | K | L | M | N | O |
|---|---|---|---|---|---|---|---|---|---|---|---|
| 11 | | | | | | | | | | | |
| 12 | | | | | | | | | | | |
| 13 | | | | | | | | | | | |
| 14 | | | | | | | | | | | |
| 15 | | | | | | | | | | | |
| 16 | 46 | | 32.8 | | 16.8 | | 0 | | 0 | | 0 |
| 17 | | | | | | | | | | | |

**FIGURE 4.7**
One row of American put option tree

Next, copy this formula once and paste it in several locations on line 16, in columns E, G, I, M, and O. If you use the same parameters that were chosen in Section 4.2, line 16 of your spreadsheet will look as in Figure 4.7.

The lines above line 16 will have entries corresponding to a European put. It is easy to convert these entries to American put values. Select any cell above, such as cell J15. Modify the formula on the special line by editing it so that it will be a MAX of the old chaining formula and the comparison value, A2 - J5. The latter value is the profit if we exercise early. The appropriate formula is

$$= \text{MAX} (\$A\$11*(\$A\$13*I16 + \$A\$12*K16) , \$A\$2 - J5)$$

Next, copy this formula once, and paste it throughout the option tree. You will see certain node entries increase as you paste, and the topmost entry will be 5 when the tree has been filled out. This is all that is needed to handle early exercise.

## EXERCISES

1. Use the spreadsheet "Amer Put" to find a put price given that the stock price at $t = 0$ is $125, the exercise price is $120, expiration is 5 periods into the future, the interest rate per period is 6.5%, $u = 1.3$, and $d = 0.96$. Identify any node where early exercise is desirable.

2. Use the spreadsheet "Amer Put" to find a put price given that the stock price at $t = 0$ is $115, the exercise price is $100, expiration is **3** periods into the future, the interest rate per period is 6%, $u = 1.1$, and $d = 0.6$. Identify any node where early exercise is desirable.

3. Use the spreadsheet "Amer Put" to find a put price given that the stock price at $t = 0$ is $125, the exercise price is $120, expiration is 5 periods into the future, the interest rate per period is 1%, $u = 1.05$, and $d = 0.96$. Identify any node where early exercise is desirable.

4. Use the spreadsheet "Amer Put" to find an *American call price*. To convert "Amer Put" to "Amer Call," edit cell K16 by changing $A6 - K6 to **K6 - $A$6**. Copy and paste this to nodes in row 16. Then edit the formula in cell J15 the same way. Copy and paste the new J15 formula in the option tree. Now rework Exercise 1 of section 4.2. Your answer is the price of an American call. How does the new answer compare with the old answer?

|   | A | B | C | D | E | F | G | H | I | J | K | L | M |
|---|---|---|---|---|---|---|---|---|---|---|---|---|---|
| 1 | 100 | Stock Price | | | | | | | | 100 | | | |
| 2 | 105 | Strike Price | | | | | | | 90 | | 110 | | |
| 3 | 5 | # Periods | | | | | | 81 | | 99 | | 121 | |
| 4 | | | | | | | 72.9 | | 89.1 | | 109 | | 133 |
| 5 | 0.05 | Interest | | | | | 65.6 | | 80.2 | | 98 | | 120 |
| 6 | 1.1 | u value | | | | | | | | | | | |
| 7 | .9 | d value | | | | | | | | | | | |
| 8 | 0.76 | q value | | | | | | | | | | | |

**FIGURE 4.8**
Stock tree

## 4.4 COMPUTING A BARRIER OPTION TREE

A European call option must be held to expiration in order to exercise it. As described in Section 3.4, this option may be modified by setting a barrier referring to the stock price (at, say, $115). If the stock price crosses this barrier at any time before expiration, the option becomes worthless. This is an example of an exotic option.

To fill in tree entries for this option price, open the computer file you saved for the European call option and save it right away with the new name, "Barrier." Look at the stock tree entries for the parameters shown in Figure 4.8.

With the barrier set at $115, the knockout option will be worth 0 at any node that is in column L or beyond. You can modify the option tree by setting all the corresponding nodes there equal to 0. This involves replacing the chaining formulas at these cells by the value 0. Once this is done, the remaining chaining formulas give the correct tree for this barrier option.

One limitation of the modification discussed above is that it works only for an initial stock price of $100. A good exercise (see the hints for Exercise 4) is to invent a formula that will automatically produce the 0 entries in the appropriate column when the initial stock price is varied. With the current stock price, it is interesting to observe the option value increasing as we enter lower strike prices.

## EXERCISES

1. Use a "Euro Call" spreadsheet to find the call price for the following barrier option. The stock price at $t = 0$ is $65, the exercise price is $70, expiration is 5 periods into the future, the interest rate per period is 5%, $u = 1.1$, and $d = 0.91$. The barrier is set at 80. If the stock rises above $80 at any time at or before expiration, the call loses all its value. Print your spreadsheet.

2. Find the call price for the barrier option in Exercise 1 if the exercise price is lowered to $55 and the barrier is lowered to $70. Print your spreadsheet.

3. Use a "Euro Call" spreadsheet to compare call prices of the following two call options. The stock price at $t = 0$ is $100, both options expire after 5 periods into the future, the interest rate per period is 4%, $u = 1.1$, and $d = 0.91$.

Option #1: A European call with an exercise price of $95.

Option #2: A *down-and-out* barrier call with the barrier set at $95. The exercise price is $95.

**4.** Find several call prices given that the stock price at $t = 0$ is $80, the exercise price is either $76 or $82, expiration is 5 periods into the future, the interest rate per period is 4.5%, $u = 1.05$, and $d = 0.88$. The call prices have an *up-and-out barrier* set at one of these values: 87, 93, or 98.

**Hint:** Zeroes can be automatically inserted in your option tree. Enter the barrier value in cell A15. Edit K16 as follows: replace "MAX..." by

$$\text{IF(K6} > \text{\$A\$15, 0, MAX ... )}$$

This is an "if" clause that compares the value in K6 with the barrier level. If it is exceeded, then 0 is used in calculations, if it is not exceeded, then the original formula is used. Edit the other option cells similarly.

## 4.5   COMPUTING *N*-STEP TREES

Where do the skills that we have developed in this chapter lead us? Option pricing trees with as few as 40 periods to as many as 100 periods give prices with enough accuracy to be useful for trading purposes. Is there some spreadsheet approach that conveniently handles the hundreds of nodes involved? Would we want to see such giant trees? Yes, there is a convenient approach; no, we want large trees to be hidden.

This chapter has served as an introduction to spreadsheet concepts, and we will build on these concepts later in this text.

# CHAPTER
# 5

## CONTINUOUS
## MODELS
## AND THE
## BLACK-SCHOLES
## FORMULA

*The race is not always to the swift, nor the battle to the strong,
but that's the way to bet it.*

Dan Parker

## 5.1  A CONTINUOUS-TIME STOCK MODEL

In the previous four chapters we studied discrete models for stock and option prices. These models are very useful from a computational point of view, and we shall return to them in Chapter 9. Now we will consider *continuous* models.

Let $S(t)$ be the price of a stock at time $t$. Our model for $S$ is given by the equation

$$dS = \mu S dt + \sigma S dB \tag{5.1}$$

where $\mu$, $\sigma$ are constants and $B$ is Brownian motion. This model for stock prices seems to have been first introduced by Paul Samuelson in 1965.[1] It is interesting to compare it with Bachelier's work in 1900.[2]

We will be working with this model at considerable length, so let us study it with care. We will first look at a discrete version of the model. The discrete version will provide us with many ideas and insights into the working of (5.1). We will then return to the continuous model and use it to derive the Black-Scholes formula.

## 5.2   THE DISCRETE MODEL

Let $S(T)$ be our stock price at time $T$. Following the spirit of the up-down model, we fix a time period $\Delta t$ and set

$$S_1 = e^{\mu \Delta t} S_0$$

and

$$S_{k+1} = e^{\mu \Delta t} S_k$$

Thus, $S_k = e^{\mu \Delta t} \cdots e^{\mu \Delta t} S_0$, where the product has $k$ factors of the form $e^{\mu \Delta t}$. This simplifies to

$$S_k = e^{\mu k \Delta t} S_0$$

Let $T = k\Delta t$ and note that

$$S_k = e^{\mu T} S_0 = S(T)$$

The last equation is saying something important. It tells us that the effect of $k$ small time steps is the same as the effect of one big time step, $T = k\Delta t$.

So far, our model is just the equation for interest compounded instantaneously. It is also the solution to the well-known differential equation

$$\frac{dS}{dt} = \mu S$$

That is,

$$S(T) = e^{\mu T} S_0 \qquad (5.2)$$

is the model we would obtain if we set $\sigma = 0$ in (5.1). This model is a deterministic one. However, stock prices are not predictable or deterministic as in (5.2). We now enrich the model.

---

[1]Samuelson, P. A. "Rational Theory of Warrant Prices," *Industrial Management Review*, 6 (1965), pp. 13–31.

[2]Bachelier, L., "Théorie de la spéculation," *Annales du Science de l'École Normale Supérieure*, 17 (1900), pp. 21–86. English translation in *The Random Character of Stock Market Prices*, ed. Cootner, P. H., Cambridge, MA: MIT Press (1964), pp. 17–78.

*Let Z denote a standard normal random variable with mean 0 and variance 1.*

The unique character of this random variable is specified by its probability density function. For each constant, $a$, the probability that $Z \leq a$ is computed with the following integral:

$$\Pr[\, Z \leq a \,] \equiv \frac{1}{\sqrt{2\pi}} \int_{-\infty}^{a} e^{-x^2/2} dx \tag{5.3}$$

We define

$$S_1 = e^{\mu\Delta t} e^{cZ_1} S_0$$

where $Z_1$ denotes one value of a standard normal, and $c$ is a constant. We can now repeat the process to generate stock prices $S_2, S_3, \cdots$ by setting

$$S_k = e^{\mu\Delta t} e^{cZ_k} S_{k-1}$$

We assume that the $Z_k$'s are independent copies of the standard normal random variable.

The repeated multiplications are summarized by the formula

$$S_k = e^{\mu k \Delta t} \exp[c(Z_1 + \cdots + Z_k)] S_0 \tag{5.4}$$

This model is more realistic, since it contains random factors. However, it suffers from one drawback, which we must fix. There are two sources of drift in (5.4). The first comes from the factor $e^{\mu\Delta t}$, which produces the drift $\mu$ for the stock. It acts like the interest rate for a bond or money market fund. The term $e^{cZ}$ produces the other source of drift. We wish to keep all the drift in the $e^{\mu\Delta t}$ factor.

We can accomplish this using an important identity for the random quantity $Z$:

$$E[e^{cZ}] = e^{c^2/2} \tag{5.5}$$

In our case, we can *normalize* our model by using equation (5.5) to correct for this unwanted drift. The identity shows that the expected value of

$$\exp[cZ - c^2/2]$$

is 1. We include this extra factor in the definition of $S_1$:

$$S_1 = e^{\mu\Delta t} \exp(cZ - c^2/2) S_0$$

The additional factor produces a random, or **stochastic**, effect that *has zero drift*, since now

$$E[S_1] = e^{\mu\Delta t} S_0$$

With this choice, our model is written as

$$S_k = S_0 e^{\mu k \Delta t} \exp[c(Z_1 + \cdots + Z_k)] e^{-kc^2/2}$$

We can make this expression more compact. The $Z_1, \cdots, Z_k$ are independent normals (mean 0, variance 1). Thus,

$$W_k = Z_1 + \cdots + Z_k$$

is a normal random variable with mean 0 and variance $k$. Our model is now expressed as

$$S(T) = S_0 e^{\mu k \Delta t} e^{c W_k} e^{-k c^2/2} \tag{5.6}$$

This may look a bit intimidating, so let us go over the factors.

- $S_0$ is simply the initial price of the stock ($t = 0$).
- $e^{\mu k \Delta t}$ is the drift or deterministic component, (compound interest factor).
- $e^{c W_k}$ is the random (stochastic) factor.
- $e^{-k c^2/2}$ is a correction factor.

***Observation.*** This discrete model is richer than our up-down model in that the rule

$$S_1 = e^{\mu \Delta t} \exp(c W_1 - c^2/2) S_0$$

permits $S_1$ to take on any positive value. Since $\exp(c W_1 - c^2/2)$ is driftless (has mean 1), one expects $S_1$ to approximate $e^{\mu \Delta t} S_0$, but it could, with small probability, be as large as $10^6 e^{\mu \Delta t} S_0$. It is easy to simulate typical values and obtain concrete paths (see Exercises 2, 3, and 4). One can set

$$S_0 = 1, \quad \mu = 0.10, \quad c = 0.40, \quad \Delta t = 1$$

and run the values for $k = 1, \dots, 42$ on a spreadsheet. A simulation selects the $Z_j$'s from the normal distribution at each step. We would obtain paths as shown in Figure 5.1.

We will make one more change to (5.6). We split time up into small increments, $\Delta t$, and take $k$ factors. So a fixed time $T = k \Delta t$ can be split into a few or possibly many increments $\Delta t$. Some consistency is needed for different choices of $k$ to refer to the same large time $T$.

The variance of $W_k$, however, increases as $k$ increases. Since $W_k$ has variance $k$, the larger $k$ is, the greater the wiggle from $W_k$. We should make sure that the total variance of $c W_k$ does not depend on $k$. This requires that the constant $c$ be adjusted with $k$.

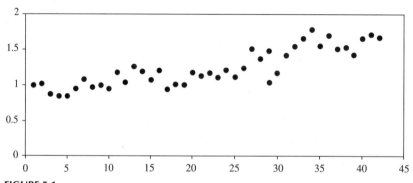

**FIGURE 5.1**
A log normal path

We choose a relationship between $c$ and $k$ so that the variance of $cW_k$ equals $\sigma^2 T$ ($\sigma$ will be a fixed model parameter). We require that

$$
\begin{aligned}
\mathrm{Var}(cW_k) &= c^2 \mathrm{Var}(W_k) \\
&= c^2 \cdot k = \sigma^2 T
\end{aligned}
$$

This allows us to substitute for $c^2 k$ in (5.6) and write our model as

$$
S_T = S_0 e^{\mu T} e^{\sigma W_T} e^{-\sigma^2 T/2}
$$

After regrouping terms, we obtain our stock model:

### Log Normal Model

$$
S_T = S_0 e^{\sigma W_T + (\mu - \sigma^2/2)T} \tag{5.7}
$$

Remember that $W_T$ is a normal random variable with mean 0 and variance $T$. Our log normal model contains two parameters, $\mu$ and $\sigma$. Let us look at several examples to see how these parameters influence a stock price.

Figure 5.2 has two simulations of the log normal model. Notice the absence of a *trend* in both graphs. This is due to the choice $\mu = 0$. The second graph appears more erratic than the first.

Now we introduce some deterministic growth. Figure 5.3 has two simulations—the first with a weak trend—and the second with a highly visible growth pattern. When $\mu$ is large compared to $\sigma$, a strong growth trend appears even though the erratic features of the path change each time we simulate the $Z$ values.

More information is visible if we plot the logarithms of the values used in the graphs in Figure 5.3. Most Internet financial services chart stock prices using a linear scale (DLJdirect, ETrade, Bigcharts), but Yahoo and Stockmaster use a log scale for their charts.

The graphs in Figure 5.4 show a small, **linear** trend for price movements corresponding to a small deterministic growth and a large, nearly linear pattern corresponding to a large growth rate. The reader should study these examples until they become familiar. They are crucial to understanding our stock model.

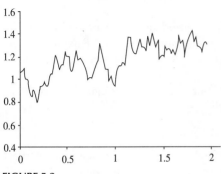 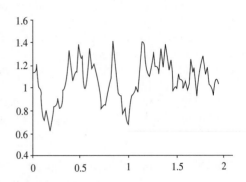

**FIGURE 5.2**
$\mu = 0$, $\sigma = 0.5$, and $\sigma = 1$

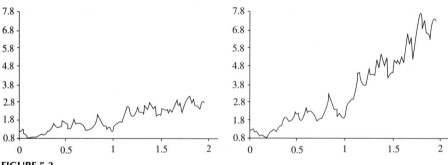

**FIGURE 5.3**

$\mu = 0.5$ and $\mu = 1$, $\sigma = 1$

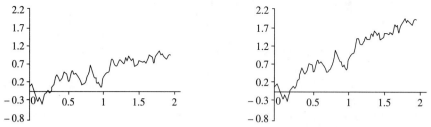

**FIGURE 5.4**

logarithmic plots: $\mu = 0.5$ (left) and $\mu = 1$ (right)

Is equation (5.7) a reasonable model for stock prices? It seems too simple to capture all the financial, economic, political, and global factors that contribute to stock price change.

Let us look at an example. We first take logarithms of both sides of equation (5.7)

$$\ln S(T) = \ln S_0 + \left(\mu - \frac{\sigma^2}{2}\right)T + \sigma W_T$$

The expression $\ln S_0 + (\mu - \sigma^2/2)T$ is just the formula for a straight line, and the term $\sigma W_T$ jiggles the points about the line. So, if we graph stock prices on a log scale (log paper), we should see them falling along a line with some random scatters off the line, if the model is realistic.

Figure 5.5 is a chart for TeleBras for the period January 10 to February 20, 1998. The graph follows a straight line with impressive regularity. This fit is satisfactory for a six-week period. So, for at least one stock over one period the model fares very well indeed.

Now that we have crafted and massaged our discrete model into the form (5.7), we will turn back to the starting point, namely (5.1):

$$ds = \mu S dt + \sigma S dB$$

The Brownian motion term is imposing. In Section 5.9 we show how to construct Brownian paths. However, there is a way to investigate (5.1) that avoids the complexities of Brownian motion.

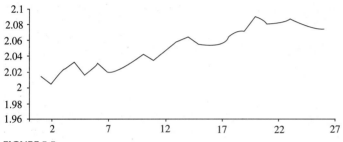

**FIGURE 5.5**
Log plot of TeleBras stock

## 5.3 AN ANALYSIS
## OF THE CONTINUOUS MODEL

Equation (5.1) is a stochastic differential equation (SDE). Most SDEs do not admit tidy closed-form solutions, but luckily this one does. That is, one can find a stochastic process whose differential, in the appropriate sense, satisfies the above SDE. This is an extraordinary stroke of good fortune.

The solution is

### Geometric Brownian Motion

$$S_t = S_0 \exp[\sigma B_t + (\mu - \sigma^2/2)t] \qquad (5.8)$$

We have seen this solution before! It is just (5.7), our discrete model with the continuous time variable $T$. Note that we did not solve the SDE. We merely asserted that (5.8) was its solution.

Here $B_t$ is a normal random variable with mean 0 and variance $t$. This gives us a model for stock prices, known as the *geometric Brownian motion* (**GBM**) model for stock prices.

Notice that

$$\ln\left(\frac{S_t}{S_0}\right) = \sigma B_t + \left(\mu - \frac{\sigma^2}{2}\right)t$$

The right-hand expression above is a normal random variable whose mean is $(\mu - \sigma^2/2)t$ and whose variance is $\sigma^2 t$.

We will use a related model to price a European call option. In order to do this, we must calibrate our stock model. There are two parameters in our model: the volatility $\sigma$ and the drift $\mu$. It turns out that to calculate the price of a European call option we need only estimate $\sigma$. The drift does not show up in the call option pricing formula. Nevertheless, we will estimate both drift and volatility using observed stock prices.

### Calibrating Geometric Brownian Motion

Suppose we have stock price data recorded during a lengthy time interval, $[0, T]$. Think of this interval as a collection of $n$ subintervals of equal length, $\Delta t$.

Assume we know the stock price at the end of each subinterval. Our data are denoted by

$$S_i : \text{Stock price at the end of the } i\text{th interval}$$

and we have $n + 1$ observations.

**Step 1** Form the data series

$$U_i = \ln(S_{i+1}) - \ln(S_i)$$

$U_1, U_2, \ldots, U_n$ is a numerical sequence, but, *before we collect our data*, the GBM model allows us to express *possible values* for $U_i$ as

$$U_i = \sigma B_{t_{i+1}} - \sigma B_{t_i} + (\mu - \sigma^2/2)\Delta t \qquad (5.9)$$

Brownian random quantities $B_{t_{i+1}} - B_{t_i}$ have these properties:

- $B_{t_{i+1}} - B_{t_i}$ is a normal random variable; its variance is $\Delta t$ and its mean is zero.
- These differences are independent random variables.

**Step 2** Find the mean and variance of the data series

$$U_1, U_2, \ldots, U_n$$

The mean is denoted by $\bar{U}$:

$$\bar{U} = n^{-1} \sum_{i=1}^{n} U_i$$

The sample variance is denoted by $S^2$:

$$S^2 = (n-1)^{-1} \sum_{i=1}^{n} (U_i - \bar{U})^2$$

These statistics are estimates for the theoretical $U$ population mean and variance. Equation (5.9) shows that all possible observations of $U$ values have mean and variance equal to

$$(\mu - \sigma^2/2)\Delta t \qquad \text{and} \qquad \sigma^2 \Delta t$$

**Step 3** Solve the equations

$$\bar{U} = (\mu - \sigma^2/2)\Delta t$$

and

$$S^2 = \sigma^2 \Delta t$$

for $\mu$ and $\sigma$. The algebra is easy. The answers are

$$\mu = \frac{\bar{U} + S^2/2}{\Delta t} \qquad \text{and} \qquad \sigma = S/\sqrt{\Delta t}$$

**Example** Daily closing prices of IBM stock from October 28, 1997, to December 9, 1997, are listed below. Two commas with no numbers between (,,) indicate that the NYSE was closed the following day.

> 99.375, 98.25, 95.812, 98.5, , 101.625, 101.938, 102.75, 101.062,
>
> 99.5, , 97.688, 99, 96.625, 99.125, 101.5, 99.125, 101.5, , 103.5,
>
> 102.125, 103.062, 104.75, 105.562, , 103.125, 107.375, 109.75,
>
> 109.5, , 112.562, 110.75, 110.375, 109.25, 112.25, , 113.062, 110.375

There are 32 prices. To estimate $\mu$ and $\sigma$ we use a spreadsheet with these prices as column A entries, in A2 to A33. Next, we form a column of price logarithms, column B. Enter the formula

$$=LN(A2)$$

in cell B2. The displayed result is 4.59890057. Select the column B cells, beginning with B2. Under the Edit menu, select Fill Down. The log formula will be copied to all the selected cells.

To complete step one, create $U_i$ values in column C. Enter the formula

$$=B3-B2$$

in cell C2. The displayed result is $-0.0113853$. Select the column C cells, and, under the Edit menu, select Fill Down. The difference formula will be copied to all the cells.

Some cells in column C have to be erased because they have entries that correspond to the empty (',,') dates; they do not have the proper $\Delta t$. We erase them by selecting a particular cell and by choosing the Clear command under the Edit menu. Cell C33 should be erased too.

Step two takes only a moment. If you are using Microsoft®Excel, obtain the mean value of column C by entering the formula

$$=AVERAGE(C2:C33)$$

in any blank cell. The spreadsheet averages the $U_i$ values in column **C**, *and ignores the blank cells*. To obtain $S$, enter the formula

$$=STDEV(C2:C33)$$

in any blank cell. The answers we get are 0.00264441 for $\bar{U}$ and 0.020256795 for $S$.

In step three, we will estimate $\mu$ and $\sigma$ on a yearly scale. One trading day corresponds to

$$\Delta t = \frac{1}{365}$$

Then our estimates for the parameters are

$$\hat{\mu} = \frac{\bar{U} + S^2/2}{\Delta t} = 1.04 \quad \text{and} \quad \hat{\sigma} = S/\sqrt{\Delta t} = 0.387$$

## EXERCISES

1. In the calibration example, $\mu$ and $\sigma$ were expressed in yearly amounts. Use the same three-step method to find these parameter values per month, then per week, then per day. What pattern do you see in your answers?

2. Use an Excel spreadsheet to list 30 values of a random $Z$ variable corresponding to equation (5.3). Then find the mean value and the standard deviation of the list, and print the spreadsheet.

*Hint:* To list values in column A, rows 2 to 31, pull down the Tools menu. Select the item, Data Analysis. The dialog box has a list of topics; scroll through the topics and select Random Number Generation. Open this topic. The new dialog box has a choice for distribution. Tap the scroll arrows to find the choice, Normal. Then, under output options, push the Output Range button. Type in the range, A2:A31. The list of random choices will appear on the spreadsheet.

3. Use an Excel spreadsheet to list 30 values of the random $W_k$ variables that appear in equation (5.6). Make an XY (scatter) chart of the list, and print the spreadsheet.

*Hint:* To list values in column B, rows 2 to 31, use the values in column A from Exercise 2. In cell B3, enter the formula =A3+B2. Then, select this cell, B3, and the cells below it and choose Fill Down from the Edit menu. Larger numbers tend to appear at the bottom of your list because a B entry is added to an A entry to give the following B entry. As the last step, enter =A2 in the cell B2. This begins column B with a $W_1$ value. Select the cells B2-B33. Then choose Chart in the menu Insert.

4. Use an Excel spreadsheet to list 30 values of the log normal model corresponding to equation(5.6). Use $\mu = 0.2$, $c = 0.1$, and $\Delta t = 1$. Make an XY (scatter) chart of the list, and print the spreadsheet.

*Hint:* List the log normal values in column D, rows 2 to 31. Use the $W_k$ values from Exercise 3. These values are in column B. In column C, place multiples of $\mu - c^2/2$. Enter this numerical value in cell C2. Enter the formula =C2+$C$2 in cell C3. Cell C3 will be twice as large as C2. Select cells C3 through C33 and choose Fill Down. You should see multiple values of the first entry. Place a log normal value in D2 by entering the formula =EXP(.1*B2+C2). Choose Fill Down to complete the column.

## 5.4   THE BLACK-SCHOLES FORMULA

The GBM stock price model (5.8) leads to a formula for the price of a European call option. This result was discovered by F. Black and M. Scholes and bears their names. We present their formula in this section, but we postpone the derivation until the next section. We are given a stock whose price today is $S_0$. Let $V$ be the value of a call option on this stock with

$$X = \text{strike price}$$
$$\tau = \text{time to expiration}$$
$$\sigma = \text{volatility of stock (constant)}$$
$$\mu = \text{drift of stock (constant)}$$
$$r = \text{riskless interest rate (constant)}$$

Then the value, $V$, of the *call* today is given by the

**Black-Scholes Formula**

$$V = S_0 N(d_1) - Xe^{-r\tau} N(d_2) \tag{5.10}$$

In this formula, $N(x)$ denotes the standard normal distribution function. That is,

$$N(x) = P[Z \leq x]$$

This distribution function is evaluated at the two points

$$d_1 = \frac{\ln(S_0/X) + (r + \sigma^2/2)\tau}{\sqrt{\tau}\sigma}$$

and

$$d_2 = d_1 - \sigma\sqrt{\tau}$$

Note that the drift rate, $\mu$, does not appear in the formula. At first glance, this seems surprising. Let us look at a numerical example to gain some feel and insight into the formula.

---

**Example**  We take Intel® on May 22, 1998, when

$$S_0 = 74.625$$
$$X = 100$$
$$\tau = 1.646 \text{ years (expiration is January 2000)}$$
$$r = 0.05$$
$$\sigma = 0.375$$

Then

$$d_1 = -0.207 \text{ and } d_2 = -0.688. N(d_1) = 0.4164 \text{ and } N(d_2) = 0.2451$$

Finally,

$$V = S_0 N(d_1) - e^{-r\tau} X N(d_2) = 8.37.$$

Actually, the press was reporting that an antitrust suit against Intel was imminent, and the market price of this call was $8.25. Bear in mind that actively traded calls such as Intel (open interest on this call = 14,410) are priced via an auction market (not by an academic formula).

---

## How to Determine Option Prices via Maple®

- Log in to Maple.
- At the > prompt, type **with(finance):** and press Enter.
- Type **evalf(blackscholes $(S, X, r, \tau, \sigma)$);** and press Enter.

where

$$S = \text{stock price}$$
$$X = \text{strike price}$$
$$r = \text{risk-free rate}$$
$$\tau = \text{time to expiration in years}$$
$$\sigma = \text{volatility}$$

Maple should display $V$, the call price.

**Remark**  Calls with lengthy expiration times are called LEAPS. They are listed in *The Wall Street Journal* and *Barron's* under long-term options.

## EXERCISES

**1.** Given the following data, compute the price of the associated European call option

|     | $S_0$ | $X$ | $r$ | $T$ | $\sigma$ |
|-----|-------|-----|-------|-------|------|
| (a) | 80  | 70  | 0.05  | 3 mo  | 0.30 |
| (b) | 60  | 66  | 0.06  | 2 mo  | 0.40 |
| (c) | 50  | 60  | 0.04  | 1 yr  | 0.25 |
| (d) | 100 | 100 | 0.055 | 4 mo  | 0.50 |
| (e) | 120 | 130 | 0.059 | 6 mo  | 0.22 |
| (f) | 40  | 40  | 0.048 | 2 yrs | 0.60 |
| (g) | 12  | 10  | 0.045 | 4 mo  | 0.33 |
| (h) | 25  | 25  | 0.05  | 3 mo  | 0.28 |

**2.** Using the data in Exercise 1, compute the price of the associated European put option for (a) through (h).

## 5.5   DERIVATION OF THE BLACK-SCHOLES FORMULA

In this section, we present a derivation of the Black-Scholes formula. The reader will be pleasantly surprised to discover how remarkably simple this argument turns out to be. It depends on a few definitions, a little bit of algebra, and a change of variable. Most of the argument amounts to identifying expressions.

### 5.5.1   The Related Model

This section may be summarized in eight words: "Make the market price = the model price." In slightly more detail, we justify the use of a "new" model (5.16). The justification is based on those eight words. Now for the details.

First, we need a stock model whose probabilities allow us to use the chaining method of Chapter 2, Section 3.3.3. To search for these probabilities, let's review some steps in Section 2.3.2.

We start with a simple investment in stock and cash. Suppose we buy $a$ shares of a stock, whose price is $S_0$, and add $b$ dollars. Our investment value is

$$\Pi = aS_0 + b \qquad (5.11)$$

Now, at some later time $\tau$, our new investment value is

$$\Pi_\tau = aS_\tau + be^{r\tau}$$

We *discount* this value, using the riskless rate, $r$,

$$e^{-r\tau}\Pi_\tau = ae^{-r\tau}S_\tau + b$$

and eliminate the $b$ term by solving (5.11). The result is

$$e^{-r\tau}\Pi_\tau = ae^{-r\tau}S_\tau + \Pi_0 - aS_0$$

After we collect terms with $\Pi$ on the left side and terms with $S$ on the right, we have a curious relation between $\Pi$ and $S$ values:

$$e^{-r\tau}\Pi_\tau - \Pi_0 = a(e^{-r\tau}S_\tau - S_0) \qquad (5.12)$$

This identity shows that $\Pi_\tau$ will inherit a property from $S$ *if we force the S model to satisfy*

$$E[\, e^{-r\tau}S_\tau - S_0 \,] = 0 \qquad (5.13)$$

We adopt (5.13) as our model "calibration" criterion, and reap the reward that, whatever $a$ we use,

$$E[\, e^{-r\tau}\Pi_\tau - \Pi_0 \,] = 0$$

The $a$ factors out, so its value does not matter. Just what is the reward? Since $E[\Pi_0] = \Pi_0$, we can compute $\Pi_0$ using *discounted future portfolio values*. We find that

$$\Pi_0 = e^{-r\tau}E[\, \Pi_\tau \,] \qquad (5.14)$$

It is a fact that, even if $a$ varies over time, the equation just above is preserved; the relation is valid for any replicating portfolio (see Section 2.3). Therefore, it is valid for a large class of option prices.

**Summary**
- Replace a stock model with the model $\tilde{S}_\tau$, where the volatility is the same, but where

$$\tilde{S}_0 = e^{-r\tau}E[\, \tilde{S}_\tau \,]$$

- Use this stock model to compute

$$e^{-r\tau}E[\Pi_\tau]$$

when $\Pi_t$ is any replicating portfolio.
- The answer is $\Pi_0$.

Is there some formula for $\tilde{S}_\tau$? Yes. The formula is closely related to (5.8) because it has the form

$$S_0 e^{\sigma B_\tau + m\tau} \tag{5.15}$$

The calibration criterion, as stated in the summary, is

$$S_0 = e^{-r\tau} E[\, S_0 e^{\sigma B_\tau + m\tau}\,]$$

which reduces to

$$E[\, e^{\sigma B_\tau + (m-r)\tau}\,] = 1$$

We see from equation (5.5) that $m = r - \sigma^2/2$.
    Thus, our related model is

$$\tilde{S}_\tau = S_0 e^{\sigma B_\tau + (r - \sigma^2/2)\tau} \tag{5.16}$$

By coincidence, the related model looks like a GBM stock model. However, it would be a strange model for a stock, since its growth rate is set artificially low. It is impractical for predicting the future, yet it is ideal for computing current values!

### 5.5.2    The Expected Value

In the case of our European call option, the final payoff is $(S_T - X)^+$, so equation (5.14) becomes

$$V = e^{-rT} E[(S_T - X)^+]$$

We are using the model given by equation (5.16):

$$S_T = S_0 \exp[\sigma B_T + (r - \sigma^2/2)T]$$

Now rewrite this expression. Recall that $B_T$ is a normal random variable with mean 0 and variance $T$. We may substitute $\sqrt{T}Z$ for $B_T$, where Z is the standard normal variable (mean 0, variance 1). Then

$$S_T = S_0 \exp[\sigma \sqrt{T}Z + (r - \sigma^2/2)T]$$

Hence,

$$V = e^{-rT} E[(S_0 \exp[\sigma \sqrt{T}Z + (r - \sigma^2/2)T] - X)^+],$$

and so,

$$V = \frac{e^{-rT}}{\sqrt{2\pi}} \int_{-\infty}^{\infty} \{(S_0 \exp[\sigma \sqrt{T}x + (r - \sigma^2/2)T] - X)\}^+ e^{-x^2/2} dx \tag{5.17}$$

by the basic rule for computing an expected value.

### 5.5.3    Two Integrals

We now set about trying to evaluate (5.17). It looks more imposing than it is. In fact, we will not even have to integrate anything! Let us first examine the expression within braces in (5.17). It is nonzero precisely when

$$S_0 \exp\left[\sigma\sqrt{T}x + \left(r - \frac{\sigma^2}{2}\right)T\right] - X > 0$$

Thus, we solve the equation $S_0 \exp[\sigma\sqrt{T}a + (r - \sigma^2/2)T] - X = 0$ for $a$ and find

$$a = \frac{\ln\left(\frac{X}{S_0}\right) - \left(r - \frac{\sigma^2}{2}\right)T}{\sigma\sqrt{T}}$$

This result is encouraging since this term is very similar to one that we saw in the Black-Scholes formula. Hence,

$$V = \frac{e^{-rT}}{\sqrt{2\pi}}\int_a^\infty [S_0 \exp\left[\sigma\sqrt{T}x + \left(r - \frac{\sigma^2}{2}\right)T\right] - X]e^{-x^2/2}dx$$

We break the integral into two pieces.

### The Second Term

$$\frac{1}{\sqrt{2\pi}}\int_a^\infty -Xe^{-x^2/2}dx = -X(1 - N(a))$$

$$= -XN(-a)$$

That was easy enough.

### The First Term We begin with

$$\frac{1}{\sqrt{2\pi}}\int_a^\infty S_0 \exp\left[\sigma\sqrt{T}x + \left(r - \frac{\sigma^2}{2}\right)T\right]e^{-x^2/2}dx$$

$$= \frac{1}{\sqrt{2\pi}}S_0 e^{(r-\sigma^2/2)T}\int_a^\infty e^{\sigma\sqrt{T}x}e^{-x^2/2}dx$$

To handle this integral we use one of the oldest techniques in mathematics: completing the square.

We write

$$x^2/2 - \sigma\sqrt{T}x = x^2/2 - \sigma\sqrt{T} + \sigma^2T/2 - \sigma^2T/2$$

$$= (x - \sigma\sqrt{T})^2/2 - \sigma^2T/2$$

Hence,

$$\frac{1}{\sqrt{2\pi}}\int_a^\infty e^{\sigma\sqrt{T}x - x^2/2}dx = \frac{1}{\sqrt{2\pi}}\int_a^\infty \exp\left[-\frac{(x - \sigma\sqrt{T})^2}{2} + \frac{\sigma^2T}{2}\right]dx$$

Now for a change of variable; let $y = x - \sigma \sqrt{T}$. Our last integral becomes

$$e^{\sigma^2 T/2} \frac{1}{\sqrt{2\pi}} \int_{a-\sigma\sqrt{T}}^{\infty} e^{-y^2/2} dy$$

$$= e^{\sigma^2 T/2}(1 - N(a - \sigma \sqrt{T}))$$

Thus the first term in (5.17) becomes, after canceling the $e^{\sigma^2 T}$ terms and simplification,

$$S_0 e^{rT} N(-(a - \sigma \sqrt{T}))$$

### 5.5.4 Putting the Pieces Together

Substituting and combining, we see that

$$V = e^{-rT} E[(S_T - X)^+]$$

$$= e^{-rT}[S_0 e^{rT} N(-(a - \sigma \sqrt{T})) - XN(-a)]$$

$$= S_0 N(-(a - \sigma \sqrt{T})) - Xe^{-rT} N(-a)$$

Since

$$a = \frac{\ln\left(\dfrac{X}{S_0}\right) - \left(r - \dfrac{\sigma^2}{2}\right)T}{\sigma \sqrt{T}}$$

Clearly

$$-a = \frac{\ln\left(\dfrac{S_0}{x}\right) + \left(r - \dfrac{\sigma^2}{2}\right)T}{\sigma \sqrt{T}}$$

and

$$-(a - \sigma \sqrt{T}) = \frac{\ln\left(\dfrac{S_0}{X}\right) + \left(r + \dfrac{\sigma^2}{2}\right)T}{\sigma \sqrt{T}}$$

Thus, $-a = d_2$, and $a - \sigma \sqrt{T} = d_1$, and finally

$$V = S_0 N(d_1) - e^{-rT} XN(d_2)$$

Please look over the argument again to appreciate how little is used in the form of sophisticated mathematics. Section 5.8 explains why a similar procedure for the tree stock models introduced in Chapter 3 gives approximate answers for the Black-Scholes value of an option: Binomial trees can be calibrated to approximate the GBM process.

## 5.6 PUT-CALL PARITY

A European call price is related to a European put price. Suppose we decide to sell a *covered* call. In other words, we buy one share of stock at price $S$, and sell one call for price $C$ (expiration and strike price are arbitrary). Then, being cautious and fearful that the stock might fall, we purchase a put, with price $P$, with the same strike and expiration as the call. Let us see where we are:

$$\text{Cost of position today } = S + P - C$$

Let us say the strike price for the put and call is $X$. What is the value of our position at expiration?

- If $S \geq X$, then Value $= X$. We give stock to the call buyer at $X$, and the put expires worthless.
- If $S < X$, then Value $= X$. We put the stock to the put seller, and the call expires worthless.

A miracle? No matter what happens, the value of the position at expiration is the same, namely $X$. Since we have a deterministic position, it follows that

$$(S + P - C)e^{r\tau} = X$$

Thus,

$$C - P = S - e^{-r\tau}X$$

Note that if the price differential, $C - P$, is not equal to $S - e^{-r\tau}X$, then arbitrageurs will drive it quickly to $S - e^{-r\tau}X$ by either buying or selling the position $S + P - C$.

We will use put-call parity to find the price of a European put option on a stock with the same parameters as earlier.

The put-call parity relation may be rewritten as

$$P = C - S + e^{-r\tau}X$$

Substituting the Black-Scholes formula for the price of the European call option on the stock, we get

$$P = SN(d_1) - e^{-r\tau}XN(d_2) - S + e^{-r\tau}X$$

Applying the well-known fact that $N(d_i) + N(-d_i) = 1$ for $i = 1, 2$, we see that

$$P = -SN(-d_1) + e^{-r\tau}XN(-d_2),$$

which is the Black-Scholes price for a European put option.

## EXERCISES

The following exercises use the random stock price

$$S_T = S_0 e^{(r-\sigma^2/2)T + \sigma\sqrt{T}Z}$$

This may be substituted for the model in equation (5.16).

1. Suppose that $Z$ denotes a standard normal random variable.
   (a) Verify that $E[e^{\sigma\sqrt{T}Z}] = e^{\sigma^2 T/2}$ for each $T > 0$.

*Hint:* Multiply the density function in the integral (5.3) by $e^{\sigma\sqrt{T}x}$ and integrate from $-\infty$ to $\infty$. Complete the square.

   (b) Show that $E(S_T) = e^{rT}S_0$
   
   Part (b) shows that the related stock model earns the continuously compounded riskless rate on the average.

2. Verify that $P(S_T > X) = N(d_2)$, where $d_2$ is one parameter in the Black-Scholes formula.

*Hint:* Read the proof of the Black-Scholes formula carefully.

3. In Chapter 1, we saw that

$$F_T = e^{rT}S_0$$

is the forward contract price of the stock that will be delivered at time $T$. Show that the parameters $d_1$ and $d_2$ in the Black-Scholes formula can be expressed as

$$\frac{\ln(F_T/X)}{\sigma\sqrt{\tau}} \pm \sigma\sqrt{\tau}/2$$

4. This exercise is computationally more intensive. Consider a *power call* option. This option has an extra parameter, $\alpha > 0$. The option pays the amount $S_T^\alpha$ if it is exercised on the expiration date $T$. Use the steps of Section 5.5 to find the following price for a power call:

$$\exp[(\alpha - 1)(r + \sigma^2/2)T + \alpha^2\sigma^2 T/2]S_0^\alpha N(\tilde{d}_1) - e^{-rT}XN(\tilde{d}_2)$$

where $\tilde{d}_i = \ln(e^{rT}S_0/X^{1/\alpha})/\sigma\sqrt{T} \pm \alpha\sigma\sqrt{T}/2$.

*Hint:* Proceed as in the proof of the Black-Scholes formula. According to the procedure in Section 5.5, the price of the power call is

$$e^{-rT}E[\,(S_T^\alpha - X)^+\,]$$

Evaluate this integral just as in the proof of the Black-Scholes formula. Use the fact that

$$S_T^\alpha = S_0^\alpha\exp[\alpha(r - \sigma^2/2)T + \alpha\sigma\sqrt{T}Z]$$

## 5.7  TREES AND CONTINUOUS MODELS

In Chapter 3 we used the chaining identity (3.3) to find option prices. Although the chaining method was applied to a tree model, we will see that tree methods are compatible with the Black-Scholes formula.

### 5.7.1  Binomial Probabilities

What does the chaining procedure accomplish? One applies it to the nodes of an option tree, working back from the expiration time, to find a value for the beginning node. Equation (3.3) of Chapter 3 has two ingredients. The discount factor $e^{-r\Delta t}$ is a minor feature of the chaining method. Its repeated use amounts to multiplying an answer with no exponential factor by the single, large discount factor $e^{-rT}$.

The answer, without the discount factor, is the major feature of chaining. It is a single expected value. This expected value is obtained by making a random choice for the *expiration node*. In a tree model, we determine this final node by tracking the price changes along a stock path. A single (random) number,

$$X = \text{number of upward moves}$$

tells us which node applies to the expiration time. The following two methods are equivalent for calculations:

- Use the chaining method to compute an expected value for a tree with $n$ time periods.
- Assume that $X$ is a binomial random variable, the number of successes in $n$ independent trials. The probability of success is the no-arbitrage probability, $q$, given in equation (3.2).

The second method uses the familiar *binomial probability* formula for values of $X$:

$$\Pr\{X = k\} = \frac{n!}{k!(n-k)!} q^k (1-q)^{n-k} \tag{5.18}$$

---

**Example: A Binomial Expected Value** A stock tree has three periods. The stock parameters are $S_0 = 20$, $u = 1.1$, $d = 0.9$, and $q = 0.8$. If an option pays an amount

$$(S_3 - 21)^+$$

at expiration, find its expected value.

The possible values of $S_3$ are 26.62, 21.78, 17.82, and 14.58. These correspond to the $X$ values 3, 2, 1, 0. We use these values in equation (5.18) to obtain the probabilities, 0.512, 0.374, 0.096, 0.008. Now, the option payoff amounts are 5.62, 0.78, 0, 0. The *expected option payoff* value is

$$5.62(0.512) + 0.78(0.374) + 0 = \$3.17$$

---

It becomes tedious to use the procedure in the example if the number of periods, $n$, is large. Fortunately, in this situation we may replace binomial probabilities by normal probabilities. This is a widely used approximation, known as

### DE MOIVRE'S CENTRAL LIMIT THEOREM

**Definition.** The central limit theorem states that for the purpose of computing typical values of $\Pr\{X \leq k\}$, we may substitute the random variable

$$\sqrt{nq(1-q)}Z + nq \tag{5.19}$$

for $X$. In this formula, $Z$ is the *standard normal random variable*.

It is not easy to understand why a binomial may be replaced by a normal. However, at least the coefficients $\sqrt{nq(1-q)}$ and $nq$ are reasonable choices. Recall that $Z$ has a mean value of 0 and standard deviation of 1.

- Adding the constant $nq$ to Z gives us a random quantity whose mean is $nq$. This is the mean value of X.
- Multiplying Z by $\sqrt{nq(1-q)}$ gives us a random quantity that has a same standard deviation as X.

---

**Example: A Normal Probability**  Suppose the price jumps of a certain stock are modeled with binomial probabilities. Its price will be observed six times a day for 15 days. Assume that the probability of an upward move is $p = 0.65$ and up/down moves are independent of each other. What is the probability of at least 52 up moves during the observation period?

Note that $np = 90(0.65) = 58.5$ and $\sqrt{npq} = \sqrt{90 \cdot 0.65 \cdot 0.35} = 4.52$. The central limit approximation for X is

$$X \approx 4.52Z + 58.5$$

To compute

$$\Pr(X \geq 52) \approx \Pr(4.52Z + 58.5 \geq 52)$$

we solve for Z to find $\Pr(Z \geq -1.55)$. The answer from a normal table is 0.94.

---

This example involved 90 time periods, but we used a single normal quantity to summarize the effect of the upward moves. One might expect that this approximation works for large stock and option trees with many periods.

Let us now consider a continuous-time stock model. The stock has drift $\mu$ and volatility $\sigma$. We wish to set up a binomial tree where individual time periods have a (brief) duration, $\Delta t$.

## 5.7.2   Approximation with Large Trees

Our tree records up or down moves from value S at some node to

$$Se^{\mu \Delta t + \sigma \sqrt{\Delta t}} \quad \text{if up}$$
$$Se^{\mu \Delta t - \sigma \sqrt{\Delta t}} \quad \text{if down}$$

at the next time period. For a fixed time $t$, we use $n$ to count the number of periods until time $t$: $n = t/\Delta t$. The node with the correct stock value at $t$ depends only on

$$X_n, \text{ the number of upward moves}$$

over the $n$ periods. To see this, think of multiplying all of the factors

$$e^{\mu \Delta t \pm \sigma \sqrt{\Delta t}}$$

that convert one node value into the next node value. After sorting the + and − terms, one finds that

$$S_t = S_0 \exp[X_n \cdot \mu \Delta t + X_n \cdot \sigma \sqrt{\Delta t}] \exp[(n - X_n) \cdot \mu \Delta t - (n - X_n) \cdot \sigma \sqrt{\Delta t}]$$

Where did those $n - X_n$ factors come from? They count the down moves. Let us focus on the exponent.

$$\ln S_t = \ln S_0 + n\mu\Delta t + (2X_n - n)\sigma\sqrt{\Delta t} \qquad (5.20)$$

So far, we have not mentioned any binomial probability. The most useful one in Chapter 2 was the arbitrage probability, equation (2.5):

$$q = \frac{S e^{r\Delta t} - S_d}{S_u - S_d}$$

We can substitute the up and down values from our model. Although the expression for $q$ does not appear simple, it is essentially equal to $1/2$; the moves are roughly symmetric. We are interested in the case when $S_u \approx S_d$, because $\Delta t$ will be quite small. If we carefully expand the $q$ value out to the $\sqrt{\Delta t}$ term, we find that

$$q \approx \frac{1}{2} - \sqrt{\Delta t} \cdot \frac{\mu + 1/2\sigma^2 - r}{2\sigma} \qquad (5.21)$$

This has some attractive implications for the random term

$$(2X_n - n)\sigma\sqrt{\Delta t}$$

in equation (5.20).

- **Mean Value**   It is $(2nq - n)\sigma\sqrt{\Delta t}$. But remember that $n = t/\Delta t$. So the mean is

$$(2q - 1)\sigma t/\sqrt{\Delta t}$$

The approximation to $2q - 1$ has a $\sqrt{\Delta t}$ factor, which cancels, and the mean is

$$(-\mu - 1/2\sigma^2 + r)t$$

Notice that $\Delta t$ has disappeared. This approximate answer depends only on the total elapsed time. Now, let us check the implication for $\ln S_t$ in equation (5.20). This expression has a $+\mu t$ and so the mean of $\ln S_t$ is

$$-\frac{1}{2}\sigma^2 t + rt = \left(r - \frac{\sigma^2}{2}\right)t$$

This last expression should look familiar. It tells us that

1. $\ln S_t$ (or its mean value) grows linearly (the way a bond or money market fund behaves)
2. The slope of this linear drift is just $r - \sigma^2/2$, which we recognize as the drift in the exponent of (5.16). The term $(r - \sigma^2/2)$ appears because we are using the risk-neutral probability, $q$.

- **Standard Deviation**   The standard deviation of $\ln S_t$ is $\text{SD}(2X_n) \cdot \sigma\sqrt{\Delta t}$. There is cancellation again:

$$\text{SD}(2X_n) = 2\sqrt{nq(1 - q)} \approx \sqrt{n} = \sqrt{t/\Delta t}$$

We ignored the term $2\sqrt{q(1-q)}$ because $q \approx 1/2$, so this term is 1. The standard deviation of $\ln S_t$ is

$$\sigma\sqrt{t}$$

**Summary.** In a large tree, with the choices of $S_u$, $S_d$, and $q$ indicated above, the random term $\ln S_t$ is approximately normal. For the purpose of calculation, we may set

$$\ln S_t = \sigma\sqrt{t}Z - \frac{1}{2}\sigma^2 t + rt = \left(r - \frac{\sigma^2}{2}\right)t + \sigma\sqrt{t}Z$$

Look carefully at the GBM model, in equation (5.16), and you will see that $S_t$ matches the GBM stock model *with the same parameter values.*

This may look elegant, but why do we need yet another link to our GBM model? Often an expected value calculation is too difficult to carry out with the GBM model. A large tree provides us with an alternative method.

For instance, if some option has a barrier, so that its value drops to zero when the barrier is crossed, then its expected value should ignore stock paths that cross this barrier. It is difficult to deal with this feature in the GBM setting, but we can be confident that a large tree and the chaining procedure give reasonably accurate expected values if we choose $\Delta t$ small. A rough guide for this is to choose $n = t/\Delta t$ so that $n$ is between 50 and 100.

### 5.7.3  Scaling a Tree to Match a GBM Model

In Chapters 3 and 4 we set up multistage stock trees by choosing node stock values based on factors $u$ and $d$. Equation (5.20) allows us to take

$$u = e^{\sigma\sqrt{\Delta t}} \qquad \text{and} \qquad d = e^{-\sigma\sqrt{\Delta t}} \tag{5.22}$$

Based on this choice for the node values, the corresponding branch probability is

$$q = \frac{1}{2} - \sqrt{\Delta t}\left(\frac{\sigma}{4} - \frac{r}{2\sigma}\right) \tag{5.23}$$

Notice that the node values and the branch probability depend only on the stock parameter $\sigma$, the interest rate $r$, and the brief time period, $\Delta t$. Smaller choices for $\Delta t$ will produce more accurate chaining results. These approximate the answers obtained if one were to use a GBM stock model to calculate the price of a stock derivative.

---

**Example—Calibrating a Large Tree**  Suppose that a certain stock has a yearly volatility $\sigma = 0.38$. Some stock options will expire in two months, so we need a tree model for the stock behavior throughout this period. To form a stock tree with 40 time periods, we calculate $u$, $d$, and $q$ using equations (5.22) and (5.23). We assume that the yearly short rate is 5.5%.

We first find $\Delta t$. Because $t = 2/12$,

$$\Delta t = t/40 = 0.00416$$

Next, $\sqrt{\Delta t} = 0.0645$. To use equation (5.22), we find that $\sigma\sqrt{\Delta t} = 0.0245$, so

$$u = e^{0.0245} = 1.0248 \qquad \text{and} \qquad d = 0.9758$$

To find the $q$ value, we use $\sqrt{\Delta t}$, $\sigma$, and $r = 0.055$ as follows:

$$q = 0.5 - 0.0645\left(\frac{0.38}{4} - \frac{0.055}{2(0.38)}\right) = 0.4985$$

This determines the correct nodes and branch weights of a 40-period tree that can be used, as in Chapters 3 and 4, to compute prices of options on this stock.

## EXERCISES

1. Suppose the price jumps of a certain stock are modeled with binomial probabilities. Its price is observed twice a day for 30 days. Assume that the probability of an upward move is $p = 0.75$. Using the normal distribution, find the approximate probability that at least 40 up moves occur during the observation period.

2. Suppose a certain stock is modeled with binomial probabilities. Each price changes by a factor of $e^{\pm 0.02}$. Assume that the probability of an upward move is $p = 0.6$, and we observe 100 price movements. Using the normal distribution, find the approximate probability that the stock increases at least 80% during the observation period.

3. Find large tree calibration factors, $u$ and $q$, for the following stock data:

$$r = 0.06, \sigma = 0.3, \Delta t = 0.003$$

4. Find large tree calibration factors, $u$ and $q$, for the following stock data:

$$r = 0.05, \sigma = 0.4, \Delta t = 0.003$$

5. Suppose that a stock option will expire at $t = 0.25$ (year.). The stock is known to have $\sigma = 0.5$. Find large tree calibration factors, $u$ and $q$, in the case that $r = 0.055$.

6. Use the data for exercise 3 to calculate the $q$ value using equation (2.5) of Chapter 2. How accurate is the formula (5.23) in this case?

## 5.8  THE GBM STOCK PRICE MODEL—A CAUTIONARY TALE

Our derivation of the Black-Scholes call price formula is based on Samuelson's stock price model,

$$dS = \mu S dt + \sigma S dB$$

This model seems reasonable, and we have used it without reservation. However, when you select a model, it may carry with it unforeseen consequences that may not fit comfortably into one's view of the real world.

Let us consider the likelihood of making a profit by exercising a European call option. Suppose we purchase a call on a stock $S$ with strike $X$, that expires at $T$. An innocent sounding question asks, "What is the probability the option will end up *in the money* (that is, will be profitable if exercised)"? This is not a difficult question to answer. Indeed,

$$\text{Pr[Option ends up in the money]}$$
$$= \Pr[S_T \geq X]$$
$$= \Pr[S_0 e^{(\mu - \sigma^2/2)T + \sigma\sqrt{T}Z} \geq X]$$
$$= \Pr[(\mu - \sigma^2/2)T + \sigma\sqrt{T}Z \geq \ln(X/S_0)]$$
$$= \Pr\left[Z \geq \frac{\ln(X/S_0) - (\mu - \sigma^2/2)T}{\sigma\sqrt{T}}\right]$$

We let

$$d = \frac{\ln(S_0/X) + (\mu - \sigma^2/2)T}{\sigma\sqrt{T}}$$

so that

$$\Pr[S_T \geq X] = \Pr[Z \geq -d] = N(d) \qquad (5.24)$$

So far, so good; but let us look at two examples to put the answer in perspective.

---

**Example**  *At-the-money* option.

$$S_0 = X = 60 \qquad T = 1 \text{ year}$$
$$\mu = 0.11 \qquad \sigma = 0.47$$

The numbers certainly apply to dozens of well-known stocks. The average yearly rate of return is 11%, a reasonable figure based on historical data of the 20th century, and the volatility is 47% per year. This is rather high by historical standards, but many stocks in 1999 exhibited such volatility. Notice that

$$d = 0$$

Thus, Pr[Option ends up in the money] = 0.5. A probability of 0.5 does not seem too unreasonable. At the same time, since the stock has an upward drift, we would expect the probability to be greater than 0.5.

---

Let us consider a more interesting case.

---

**Example**  Again, consider an at-the-money option, with

$$S_0 = X = 60 \quad T = 1 \text{ year}$$
$$\mu = 0.20 \quad \sigma = 1.00$$

The volatility is high, but this level is not unheard of. One finds that

$$d = -0.3$$

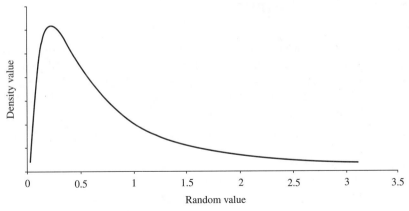

**FIGURE 5.6**
Density of GBM, $\mu = 0$

Thus,

$$Pr[\text{Option ends up in the money}] =$$

$$= N(-0.3) = 1 - 0.618 = 0.38$$

This result certainly goes against intuition. This is a stock with a yearly rate return of 20%. How can a stock with a strongly positive growth rate finish *out of the money* almost twice as often as it finishes in the money? We would expect that $Pr[S_t \geq X]$ should be at least 0.5.

The calculations are correct. The problem lies in our original model,

$$dS = \mu S dt + \sigma S dB,$$

which leads to the equation

$$S_t = S_0 \exp\left[\left(\mu - \frac{\sigma^2}{2}\right)t + \sigma B_t\right]$$

The probability distribution for the random variable

$$\exp\left[\sigma B_t - \frac{\sigma^2 t}{2}\right]$$

is very lopsided for large $\sigma$. Figure 5.6 is a graph of its density function. It is this misshapen distribution that produces the paradoxical results.

The Black-Scholes formula is much too useful to discard merely because of this minor failure to fit real-world intuition. At the same time, the reader should be aware that our simple model is not without its shortcomings. Perhaps a better model lurks in the wings.

## 5.9  APPENDIX: CONSTRUCTION OF A BROWNIAN PATH[3]

Given a fair coin, one could toss it 100 times and list the results in order of their occurrence. This list would be one of many possibilities. We will show how to construct a Brownian path $W = W(t)$ on the interval [0, 1].

Let $Z$ denote a random quantity that has a standard normal distribution. Divide the interval into 100 equal pieces. Thus we can set

$$t_j = j/100$$

for $j = 0, \ldots, 100$. One might set

$$W(t_k) = \sum_{j=1}^{k} Z(j)$$

where $Z(j)$ are random points. Set $W(0) = 0$. To get the Brownian path, connect the points $(t_k, W(t_k))$ with line segments. But this definition produces a curve that is too spikey. It could be smoothed by scaling as follows. Set

$$W(t_k) = \sum_{j=1}^{k} Z(j) \frac{1}{100}$$

Note that $\frac{1}{100} = \Delta t$. But this produces a path that hardly varies from $W(0)$. The correct scaling is given by

$$W(t_k) = \sum_{j=1}^{k} Z(j) \frac{1}{10}$$

Note that $\frac{1}{10} = \sqrt{\Delta t}$. Rephrased, we have chosen

$$\Delta = Z \cdot \sqrt{\Delta t}$$

We have produced a curve, but this is not the end of the story. Our path consists of a bunch of points joined by line segments. To get a true Brownian path, we must let $\Delta t \to 0$. We have to arrange matters carefully so that we end up with a well-defined quantity at the end.

See Figures 5.7 through 5.9 for some intermediate "Brownian paths." Figure 5.8 has more random $Z$ values than Figure 5.7, but these were not obtained by simply resimulating $Z$ values over 32 time periods. To produce a path that will converge as $\Delta t \to 0$, the points in a previous picture must be retained. That is, *intermediate* random points *must be added* in a consistent manner in order to reach a limit.

Paul Levy discovered a method to add new points consistently along a Brownian path. His procedure is based on how the midpoint of a path behaves if given

$$B_t = A \qquad \text{and} \qquad B_{t+\Delta t} = B$$

---

[3]This topic may be omitted on first reading.

**FIGURE 5.7**
16 time intervals

**FIGURE 5.8**
32 time intervals

**FIGURE 5.9**
64 time intervals

Under these conditions, the midpoint is still normal and it can be simulated using a new $Z$ value:

$$B_{t+\Delta t/2} = \frac{A + B}{2} + \frac{\sqrt{\Delta t}}{2} Z$$

---

**Example**  Suppose that $B_2 = -0.6$ and $B_4 = 1.2$. To produce a consistent value for $B_3$, we generate a $Z$ value using a random number generator. Then we set

$$B_3 = 0.3 + \frac{\sqrt{2}}{2} Z$$

After calculating $B_3$, we can continue. A value for $B_{2.5}$ can be simulated with

$$B_{2.5} = \frac{-0.6 + B_3}{2} + \frac{1}{2} Z'$$

provided that one chooses a new $Z'$ value[2].

---

The only rule for continuing this simulation is that each $Z'$ value must be independent of the previous $Z$ values. When one stops simulating points, then line segments between the points give a reasonable approximation to a Brownian path. To obtain a better approximation, add more points!

**Historical Note**  Brownian motion is a mathematical construct. It is named for Robert Brown, English biologist (1773–1858), who performed laboratory experiments to determine why bits of tissue and small particles jiggle about in random directions when viewed through a microscope. Einstein gave the correct explanation of this fact in 1905; he based his theory on collisions with molecules.

---

[2]Karlin and Taylor, *A First Course in Stochastic Processes*. New York: Academic Press, 1975.

# CHAPTER
# 6

## THE ANALYTIC APPROACH TO BLACK-SCHOLES

*Mathematicians are like Frenchmen; you tell them something, they translate it into their own language, and it comes out completely different.*

Goethe

In this chapter we will use our simple stock model,

$$dS = \mu S dt + \sigma S dB \qquad (6.1)$$

to construct an analytic model for an option price. The analytic model will be a differential equation. Differential equations have served as models in all fields of human endeavor. Their importance and usefulness was first recognized by Isaac Newton, who stated the basic task as

> Data aequatione quotcunque fluentes quantitates involvente fluxiones invenire et vice versa

Three centuries later the task is still the same. Wherever in the world there are quantities that change over time (Newton's "*fluentes quantitates*"), we can find differential equations (Newton's "*fluxiones*") that model the situation and then solve those differential equations for the changed values, whether they are populations, temperatures inside stars, or option prices. Thus, differential equations provide a powerful tool for modeling the physical world.

## 6.1  STRATEGY FOR OBTAINING THE DIFFERENTIAL EQUATION

Let $V(S_t, t)$ denote the price of an option on a stock: $S_t$ is the price of the stock at time $t$. We assume that $V(S, t)$ is a smooth function of (generic) variables $S$ and $t$. The strategy for constructing the Black-Scholes differential equation has four steps. The first three steps are purely mathematical, and the crucial financial argument is provided in the fourth step.

> **Step 1**  Expand $V$ in a Taylor series in $S$ and $t$.
>
> **Step 2**  Substitute the expression for $dS$, equation (6.1), in the Taylor series for $V$.
>
> **Step 3**  Perform algebraic simplifications, simplify the Brownian term, and discard higher-order terms.
>
> **Step 4**  Equate $V$ to a replicating portfolio.

We then arrive at the desired differential equation.

## 6.2  EXPANDING V(S, t)

Let $y = f(x)$. Recall that the Taylor series expansion for $f$ has the following form

$$f(x) \sim \sum_{k=0}^{\infty} \frac{f^k(0)}{k!} x^k$$

in the one-variable case. We are interested in the two-variable case, $z = f(x, y)$, where the Taylor series has the form

$$f(x, y) = f(0, 0) + \frac{\partial f(0, 0)}{\partial x} x + \frac{\partial f}{\partial y}(0, 0)y + \frac{1}{2} \frac{\partial^2 f(0, 0)}{\partial x^2} x^2 \qquad (6.2)$$

$$+ \frac{\partial^2 f(0, 0)}{\partial x \partial y} xy + \frac{1}{2} \frac{\partial^2 f(0, 0)}{\partial y^2} y^2 + \text{higher-order terms}$$

Let us rewrite (6.2) in differential form

$$df = \frac{\partial f}{\partial x} dx + \frac{\partial f}{\partial y} dy + \frac{1}{2} \frac{\partial^2 f}{\partial x^2} dx^2$$

$$+ \frac{\partial f}{\partial x \partial y} dx dy + \frac{1}{2} \frac{\partial^2 f}{\partial y^2} + \text{higher order terms} \dots \qquad (6.3)$$

You may interpret $dx$ as a small change in $x$ and $df$ as a small change in $f$ that results from the small changes $dx$, $dy$. We have written equation (6.3) for $f$, so that you may see the general mathematical form, independent of the field (heat, light, sound, electromagnetism).

Now apply the expansion to $V$:

$$dV = \frac{\partial V}{\partial S}dS + \frac{\partial V}{\partial t}dt + \frac{1}{2}\frac{\partial^2 V}{\partial S^2}(dS)^2$$

$$+ \frac{\partial^2 V}{\partial S \partial t}dSdt + \frac{1}{2}\frac{\partial^2 V}{\partial t^2}(dt)^2 + \text{ higher order terms}$$

This completes Step 1.

## 6.3   EXPANDING AND SIMPLIFYING $V(S_t,t)$

We will now substitute (6.1) in the above equation for $dS$ and keep only first-order terms in $dt$.

$$dV = \frac{\partial V}{\partial t}dt + \frac{\partial V}{\partial S}(\mu Sdt + \sigma SdB) + \frac{1}{2}\frac{\partial^2 V}{\partial S^2}(\mu Sdt + \sigma SdB)^2$$

$$+ \frac{\partial^2 V}{\partial S \partial t}(\mu Sdt + \sigma SdB)dt + \frac{1}{2}\frac{\partial^2 V}{\partial t^2}dt^2$$

We keep the first four terms and discard the last term on the right-hand side. We now must look carefully at what remains.

$$dSdt = (\mu Sdt + \sigma SdB)dt = \mu S(dt)^2 + \sigma SdBdt$$

Recall that $dB \simeq Z\sqrt{dt}$, so that $dBdt \simeq Z(dt)^{3/2}$. Since both $(dt)^2$ and $(dt)^{3/2}$ are not first order, we discard $dSdt$. This leaves only $(dS)^2$ to consider:

$$(dS)^2 = \mu^2 S^2 (dt)^2 + \mu\sigma S^2 dBdt + \sigma^2 S^2 (dB)^2$$

We discard the first two terms on the right by the previous arguments. On the other hand, $(dB)^2 \simeq (Z\sqrt{dt})^2 = Z^2 dt$, which is first order in $dt$! We retain this term.

Now, assemble the pieces for $dV$:

$$dV = \frac{\partial V}{\partial t}dt + \frac{\partial V}{\partial S}(\mu Sdt + \sigma SdB) + \frac{1}{2}\sigma^2 S^2 Z^2 \frac{\partial^2 V}{\partial S^2}dt$$

$$= \left(\frac{\partial V}{\partial t} + \mu S\frac{\partial V}{\partial S} + \frac{1}{2}\sigma^2 S^2 Z^2 \frac{\partial^2 V}{\partial S^2}\right)dt + \sigma S\frac{\partial V}{\partial S}dB \qquad (6.4)$$

To simplify (6.4) further, we will replace $Z^2$ by its expected value, namely 1. We thus obtain

$$dV = \left(\frac{\partial V}{\partial t} + \mu S\frac{\partial V}{\partial S} + \frac{1}{2}\sigma^2 S^2 \frac{\partial^2 V}{\partial S^2}\right)dt + \sigma S\frac{\partial V}{\partial S}dB \qquad (6.5)$$

The justification for $Z^2 \to 1$ is that any sum of these differentials contains an average that nearly equals the mean value. We have made the choice specified above for clarity and transparency.

This completes Steps 2 and 3. The steps thus far have been completely mathematical. $V$ could have been any physical quantity. To complete our program we need a financial tool.

## 6.4  FINDING A PORTFOLIO

In Chapter 2, Section 2.3, we developed a very powerful tool, the replicating portfolio. This allowed us to compute derivative prices for certain discrete time models. In the present context, the same tool will help us. We will take advantage of some technical facts concerning the stock model given in equation (6.1).

This knowledge allows us to assess an overall investment return as a stock position varies over time. We begin a search for a suitable *stock/bond* investment. Our goal will be that, at any time, the net worth of this investment is the target value, $V(S_t, t)$.

We are assuming, as before, that $V(S, t)$ is some given, smooth function of the variables $S$ and $t$. We will find numerical values

$$\phi(t) = \text{number of shares of stock}$$

and

$$\psi(t) = \text{number of bonds}$$

so that the equation

$$V(S_t, t) = \phi S_t + \psi P_t \tag{6.6}$$

holds for $0 \le t \le T$. This equation states that the net worth of our position is merely the value of our stock and bond holdings. $P_t$ is the value of a bond. We assert that changes in the net worth obey the equation

$$dV = \phi dS_t + \psi dP_t \tag{6.7}$$

This assertion involves a subtle point, and we will present a careful and detailed explanation for (6.7) in Section 6.7. At present, we wish to complete the derivation of the Black-Scholes equation.

Since

$$dS = \mu S dt + \sigma S dB$$

and

$$dP = rP dt$$

we can rewrite (6.7) as

$$dV = (\mu \phi S + r \psi P)dt + \sigma \phi S dB$$

We next equate the $dV$ from (6.7) and (6.5):

$$(\mu \phi S + r \psi P)dt + \sigma \phi S dB = \left( \frac{\partial V}{\partial t} + \mu S \frac{\partial V}{\partial S} + \frac{1}{2}\sigma^2 S^2 \frac{\partial^2 V}{\partial S^2} \right) dt + \sigma S \frac{\partial V}{\partial S} dB$$

This looks grim, but help is on the way. We have considerable latitude in the choice of $\phi$ and $\psi$, so we set

$$\phi(t) = \frac{\partial V}{\partial S}(S_t, t) \tag{6.8}$$

With this choice, not only do the $dB$ terms cancel, but $\mu\phi S$ and $\mu S\partial V/\partial S$ cancel also. This leaves

$$r\psi P dt = \left(\frac{\partial V}{\partial t} + \frac{1}{2}\sigma^2 S^2 \frac{\partial^2 V}{\partial S^2}\right)dt \qquad (6.9)$$

We see that $\psi P = V - S\partial V/\partial S$ from equation (6.6). Substitute this in (6.8) to obtain

$$r\left(V - S\frac{\partial V}{\partial S}\right)dt = \left(\frac{\partial V}{\partial t} + \frac{1}{2}\sigma^2 S^2 \frac{\partial^2 V}{\partial S^2}\right)dt$$

We arrive at the equation

### Option Price PDE

$$\frac{\partial V}{\partial t} + \frac{1}{2}\sigma^2 S^2 \frac{\partial^2 V}{\partial S^2} + rS\frac{\partial V}{\partial S} - rV = 0 \qquad (6.10)$$

This is the celebrated **Black-Scholes partial differential equation** for the price of a stock option. Equation (6.10) represents an important breakthrough in financial thinking.

### Initial Conditions and the Black-Scholes Differential Equation

To obtain an option price, such as a European call price, equation (6.10) must be combined with *boundary conditions*. The three boundary conditions for a call are

1. The Payoff: $V(S, \tau) = (S - X)^+$
   This condition states the obvious; the payoff on the option is exactly what it should be.
2. $\lim_{S\to\infty} V(S, t)/S = 1$ for a deep-in-the-money option. The value of the option approaches $S - X$, and so the ratio approaches 1.
3. $S(t_0) = 0$ implies that $V(S, t) = 0$ for $t > t_0$. Once the stock becomes worthless, it does not recover.

   *Remark.* At first glance (6.10) might appear a bit imposing. Although we will not do so, we could change variables in a clever way and reduce the equation to

$$\frac{\partial^2 V}{\partial S^2} = \frac{\partial V}{\partial t}$$

In this form, the equation appears in every book on partial differential equations.

## EXERCISES

1. Suppose $V(S, t) = aS + be^{rt}$. Verify that, for any choice of constants $a$ and $b$, $V$ satisfies the Black-Scholes equation (6.10). Analyze the boundary behavior of $V$ at expiration and under the conditions $S = 0$ and $S \to \infty$. What financial derivative does $V$ represent if $a = 1$ and $b < 0$?

**2.** Suppose $V(S, t) = e^{at}S^2$ . Verify that, for a certain choice of $a$, $V$ satisfies the Black-Scholes equation.

*Hint:* The answer is $a = -(\sigma^2 + r)$.

**3.** Suppose $G(S, t)$ solves the following equation:

$$\frac{\partial G}{\partial t} + \frac{1}{2}\sigma^2 S^2 \frac{\partial^2 G}{\partial S^2} + rS\frac{\partial G}{\partial S} = 0$$

This is not the Black-Scholes equation, because it is missing the final term, $-rG$. Verify that

$$V(S, t) = e^{rt}G(S, t)$$

does satisfy the Black–Scholes equation.

## 6.5 SOLVING THE BLACK-SCHOLES DIFFERENTIAL EQUATION

One of the easiest ways to solve a differential equation is to guess a solution and try it. So, we "guess" [see equation (5.10)]

$$V = S_0 N(d_1) - Xe^{-rT}N(d_2) \tag{6.11}$$

substitute it into the differential equation, and verify that it does indeed satisfy equation (6.10). We leave the calculations to the reader, but this process does confirm that (6.11) is a solution.

There are many other solutions to equation (6.10). In order to calculate an option price, we find a solution that agrees with the option's value on its expiration date. Since this solution of (6.10) matches *every possible* expiration value, the price formula we obtain applies to both earlier and later dates.

We illustrate this procedure for three specific stock derivatives. First, we show why the normal distribution, given by equation (5.3), generates an important solution of equation (6.10). This solution leads to prices for two other derivatives.

### 6.5.1 Cash or Nothing Option

Recall that in Section 5.2, the quantity $N(x)$ was introduced as the probability $\Pr[Z \leq x]$. Notice that $N$ has these properties:

- As $x \to +\infty$, $N \to 1$
- As $x \to -\infty$, $N \to 0$
- $N' = (1/\sqrt{2\pi})e^{-x^2/2}$, which leads to the identity $N'' = -xN'$

The first two properties help us match a solution with expiration values and boundary conditions. The identity for $N''$ will assist in verifying that (6.10) is satisfied.

Suppose that positive values $X$ and $\sigma$ are given and we choose any constant $b$. The following expression will be substituted for $x$ in $N(x)$:

$$d(t, S) = \frac{\ln(S/X)}{\sigma\sqrt{T-t}} + b\sqrt{T-t}$$

The function $d(t, S)$ has these properties:

- As $t \to T$, if $S > X$, then $d \to +\infty$
- As $t \to T$, if $S < X$, then $d \to -\infty$
- $\partial d / \partial t = d / 2(T - t) - b / \sqrt{T - t}$

We define $\tilde{V}$ as follows:

$$\tilde{V}(S, t) = N(d) \tag{6.12}$$

Equation (6.12) is almost the price of a financial derivative. Notice that from properties of $N$ and $d$, as time $t$ approaches the expiration date $T$, the value of $\tilde{V}$ approaches either one or zero. In fact, $\tilde{V} \to 1$ if $S > X$ and $\tilde{V} \to 0$ if $S < X$.

We say that this behavior at expiration represents a **cash or nothing derivative**. The derivative payoff at time $T$ is

- One dollar, if the stock is doing well $(S > X)$
- Zero dollars, if the stock is doing poorly $(S < X)$

Also, if $S \to 0$ at any time, then $d \to -\infty$, so $\tilde{V} \to 0$; the option becomes worthless. Similarly, as $S \to \infty$ then $d \to \infty$ and so $\tilde{V} \to 1$. These are correct boundary values for a cash or nothing derivative.

Is $\tilde{V}$ the price of a financial derivative? Well, almost. Verify the identity

$$\frac{\partial \tilde{V}}{\partial t} + \frac{1}{2}\sigma^2 S^2 \frac{\partial^2 \tilde{V}}{\partial S^2} + rS\frac{\partial \tilde{V}}{\partial S} = 0 \tag{6.13}$$

for $\tilde{V}$. The verification involves taking derivatives of $N(d)$ and expressing the answers in terms of $N'$ and $d$. You can see that the special choice of $b = r/\sigma - \sigma/2$ is needed for (6.13) to hold.

Unfortunately, equation (6.13) is not the Black-Scholes equation. But this problem is easily fixed. If we define $V$ by

$$V = e^{-r(T-t)}\tilde{V}$$

then $V$ still has the same boundary behavior, but the extra factor with $e^{rt}$ enables $V$ to satisfy the Black-Scholes equation (6.10). So,

$$V(t, S) = e^{-r(T-t)}N[d(t, S)] \tag{6.14}$$

is the derivative price at $t$ when we substitute the market quote for $S$.

### 6.5.2  Stock-or-Nothing Option

Equation (6.12) represents a simple option payoff. At the expiration date, we obtain a dollar or we obtain nothing. An interesting modification of this is the following:

$$V = S \cdot N(d)$$

As in the previous section, $d$ is defined to be

$$d(t, S) = \frac{\ln(S/X)}{\sigma\sqrt{T-t}} + b\sqrt{T-t}$$

We see from earlier comments that, as the expiration date approaches, the value of $V$ tends to either $S$ or zero. In fact, $V \to S$ if $S > X$ and $V \to 0$ if $S < X$. Also, if $S \to 0$ at any time, then certainly $V \to 0$, and if $S \to \infty$ then $V \approx S$.

This boundary and expiration behavior represents a **stock or nothing derivative**. The derivative payoff at time $T$ is

- one unit of stock if the stock is doing well ($S > X$)
- nothing if the stock is doing poorly ($S < X$).

Is $V$ an option price? Yes, it is. The reader should verify that equation (6.10) is satisfied by this formula. The presence of the factor $S$ changes some partial derivatives; after you combine several terms, you will find that a *new* choice for the constant $b$ is needed in order that (6.10) is satisfied. The new value is

$$\tilde{b} = r/\sigma + \sigma/2$$

We will denote the *new* $d$ function by $\tilde{d}$. Then the stock or nothing option price is simply

$$V = SN(\tilde{d}) \tag{6.15}$$

### 6.5.3 European Call

We combine the two options we have discussed in this section to obtain a call price.

Let us suppose that we *hold* one *stock or nothing* option, with the payoff level set at $X$, and *sell* $X$ units of the *cash or nothing* option. Then our net investment value at any time is the difference of prices given by equations (6.14) and (6.15):

$$SN(\tilde{d}) - Xe^{-r(T-t)}N(d)$$

Consider the expiration value of this investment. If the stock does poorly, then neither option pays off. But if the stock does well ($S > X$), then the payoff is $S - X$.

This *is the payoff for a European call*. Indeed, the expression above is the Black-Scholes price formula. We have derived it using prices of two other options.

## 6.6 OPTIONS ON FUTURES

Trades of futures contracts involve lower transaction costs than trades of the underlying asset. This makes it less expensive to adjust a portfolio containing futures than one composed only of stock. Options on futures contracts are readily bought and sold in major markets.

A European call on a futures contract is a commonly traded option. Usually the call expires on the delivery date (settlement date) for the futures contract. Since the delivery price will equal the asset price, we would expect a futures call price to be closely related to an "ordinary" call price.

Let us denote the futures price by $F$. We have seen in Chapter 1 that $F$ and $S$ are related via the equation

$$F_t = S_t e^{r(T-t)}$$

Notice that the futures price, $F$, is directly and simply related to the stock price.

## 6.6.1 Call on a Futures Contract

Let us begin with the Black-Scholes formula (5.10) for an ordinary call and express it in terms of the futures price. Equation (5.10) contains the expressions

$$d_1 = \frac{\ln(S_0/X) + (r + \sigma^2/2)\tau}{\sqrt{\tau}\sigma} \tag{6.16}$$

and

$$d_2 = d_1 - \sqrt{\tau}\sigma, \qquad \tau = T - t$$

Substituting $Fe^{-r\tau}$ for $S_0$, we obtain

$$d_1 = \frac{\ln(Fe^{-r\tau}/X) + (r + \sigma^2/2)\tau}{\sqrt{\tau}\sigma}$$

Expanding the log and canceling, we arrive at

$$d_1 = \frac{\ln(F/X) + \sigma^2\tau/2}{\sqrt{\tau}\sigma}$$

Similarly,

$$d_2 = \frac{\ln(F/X) - \sigma^2\tau/2}{\sqrt{\tau}\sigma}$$

Consequently, both $d_1$ and $d_2$ are simple functions of the futures price. Let us go further. Equation (5.10) states that a call price is

$$S_0 N(d_1) - e^{-r\tau} X N(d_2) = e^{-r\tau}[e^{r\tau} S_0 N(d_1) - X N(d_2)]$$

We see the appearance of the futures price in the Black-Scholes formula, and we write the result as

**Call Option on a Future**

$$C(F, t) = e^{-r(T-t)}[F N(d_1) - X N(d_2)] \tag{6.17}$$

In equation (6.17),

$$d_1 = \frac{\ln(F/X) + (\sigma^2/2)(T - t)}{\sigma\sqrt{T - t}}$$

and

$$d_2 = d_1 - \sigma\sqrt{T - t}$$

It turns out that we have found the value of a call on a future. Note that $d_1$ and $d_2$ are even simpler here than in the stock case. The next section will justify this formula by showing that this price can be hedged.

---

**Example**   A stock index is currently at 702, and an estimate for the volatility of this index is $\sigma = 0.4$. An index futures contract will expire in three months and will pay the index amount, at that time, in dollars. The futures contract sells for $715.

A call on this future, with the same expiration date, has an exercise price of $740. If the short rate is 7%, what is the computed price of this call option?

With our terminology,

$$F = 715 \qquad T - \tau = 0.25 \qquad \sigma = 0.4$$
$$X = 740 \qquad r = 0.07$$

We use equation (6.17):

$$F/X = 715/740 = 0.9662 \qquad \text{and} \qquad \sigma\sqrt{T - t} = 0.4(0.5) = 0.2$$
$$d_1 = \ln(0.9662)/0.2 + 0.2/2 = -0.071922$$
$$d_2 = d_1 - 0.2 = -0.271922$$

Then, from a normal table,

$$N(d_1) = 0.4721 \qquad \text{and} \qquad N(d_2) = 0.3936$$

Finally, we evaluate (6.17):

$$G = (0.98265)(0.4721 \cdot 715 - 0.3936 \cdot 740)$$
$$= \$45.48$$

---

### 6.6.2   A PDE for Options on Futures

The Black-Scholes partial differential equation can be solved to find option prices, not only on the underlying asset (the stock in this case) but also on the corresponding futures contract. The call price formula in equation (6.17) is a major example.

As before, we denote the futures price by $F$. We have seen that $F$ and $S$ are related via the equation

$$F_t = S_t e^{r(T-t)}$$

Now suppose that some option on a future has a price given by

$$G(F_t, t)$$

and the function $G(F, t)$ has partial derivatives. We will find an equation for $G$ by "pretending" that we have a stock option.

We are thinking of the price

$$G(Se^{r(T-t)}, t),$$

as a financial derivative of the stock. Consequently, this price satisfies the differential equation (6.10). Let us look at some partial derivatives. First,

$$\frac{\partial}{\partial S} G(Se^{r(T-t)}, t) = \frac{\partial G}{\partial F} e^{r(T-t)}$$

We used the chain rule here, because our price has the form $G(cS, t)$. If we differentiate again, we obtain

$$\frac{\partial^2}{\partial S^2} G(Se^{r(T-t)}, t) = \frac{\partial^2 G}{\partial F^2} e^{2r(T-t)}$$

Take a look at equation (6.10). The one term with $\partial G / \partial S$ also has a factor of $rS$. But this fits nicely with the first partial above. Since $Se^{r(T-t)} = F$, we obtain

$$rS\frac{\partial G}{\partial S} = rF\frac{\partial G}{\partial F}$$

Similarly, the one term with $\partial^2 G / \partial^2 S$ also has a factor $\sigma^2 S^2 / 2$. We can simplify to obtain

$$\frac{\sigma^2}{2} S^2 \frac{\partial^2 G}{\partial S^2} = \frac{\sigma^2}{2} F^2 \frac{\partial^2 G}{\partial F^2}$$

The only other partial derivative we need is

$$\frac{\partial}{\partial t} G(Se^{r(T-t)}, t)$$

Since $t$ appears twice in this function, we must add two partials:

$$\frac{\partial}{\partial t} G(Se^{r(T-t)}, t) = -rSe^{r(T-t)}\frac{\partial G}{\partial F} + \frac{\partial G}{\partial t}$$

The right-hand side simplifies to

$$-rF\frac{\partial G}{\partial F} + \frac{\partial G}{\partial t}$$

Notice that the $-rF \ldots$ term above *will cancel* the term with the first partial in the Black-Scholes equation. When we add all the terms for equation (6.10), we obtain a rather simple PDE:

### Options on Futures

$$\frac{\partial G}{\partial t} + \frac{1}{2}\sigma^2 F^2 \frac{\partial^2 G}{\partial F^2} - rG = 0 \tag{6.18}$$

## EXERCISES

1. Verify that the call option price in equation (6.17) does satisfy the PDE (6.18).
2. A stock index is currently at 1450, and an estimate for the volatility of this index is $\sigma = 0.2$. An index futures contract will expire in three months and will pay the index amount in dollars. The futures market price is $1472. A call on this future has an exercise price of $1460. If the interest rate is 5%, what is the computed price of this call option?

**3.** Suppose that $V(S, t)$ solves equation (6.10) *for the case* $r = 0$. Verify that the formula

$$G(F, t) = e^{-r(T-t)} V(F, t)$$

satisfies the *futures* equation (6.18). This illustrates how closely these types of option prices are related.

## 6.7  APPENDIX: PORTFOLIO DIFFERENTIALS

A continuous-time investment in a stock and bond can be modeled with the equations

$$\Pi_t = \phi S_t + \psi P_t$$

and

$$d\Pi_t = \phi dS_t + \psi dP_t$$

The second equation, (6.7) in Section 6.4, is not, in any sense, a mathematical chain rule that has gone astray. To see this, and to appreciate the importance of (6.7), we begin our discussion by assuming that some portfolio value is meaningful and varies over time.

Suppose that we arbitrarily break up some *lengthy* time interval into finitely many *brief* subintervals $[t_k, t_{k+1}]$. Then the total portfolio change can be recovered from its changes on each brief interval. It is clear that

$$\Pi_T - \Pi_0 = \sum \Delta\Pi_{t_k}$$

The expression $\Delta\Pi_{t_k}$ is an abbreviation for $\Pi_{t_{k+1}} - \Pi_{t_k}$. Next, suppose that on each brief time interval, the stock and bond positions, $\phi$ and $\psi$, *were frozen* at the beginning of each time interval.

This produces a slightly different portfolio that will have different values over the lengthy time interval. We denote these values by $F_t$.

The frozen positions allow us to express the correct return over each brief interval:

$$\Delta F_{t_k} = \phi_{t_k}\Delta S_{t_k} + \psi_{t_k}\Delta P_{t_k} \tag{6.19}$$

*The point here is that the value changes only if the asset values change.*

Our objective is to demonstrate that $F_T$ leads to an identity for $\Pi_T$. We certainly have the following relation for $F_T$:

$$F_T - F_0 = \sum \Delta F_{t_k}$$

Now we substitute from equation (6.19) to obtain an identity with $\Delta S$ and $\Delta P$.

$$F_T - F_0 = \sum \phi_{t_k}\Delta S_{t_k} + \sum \psi_{t_k}\Delta P_{t_k}$$

At this point, we can use some technical knowledge of our stock model. It is known[1] that as $\Delta t_k \to 0$, each sum above has a limit. The answer is expressed

---

[1] Karatzas, I., and Shreve, S., *Brownian Motion and Stochastic Calculus*, 2nd edition, Springer-Verlag, New York, 1996.

mathematically as

$$\int_0^T \phi(t)dS + \int_0^T \psi(t)dP \tag{6.20}$$

since each limit is a Lebesgue-Stieltjes integral.

We will not compute any such answer. It is enough to know that there is a limit in the case when $\phi$ (and consequently $\psi$) *are merely continuous*. It is important to utilize the *invariance* of this limit.

The answer is not dependent on the choice of time points used to freeze the investment choices. Every possible $F_t$ value, constructed by freezing investment positions, converges to the quantity given by equation (6.20) as the brief time intervals shrink to zero.

Is (6.20) the value of $\Pi_T$ then? Perhaps not. But we will stipulate that it is.

We use these integrals *to define* $\Pi_T$. Such an argument seems to be circular. The reason for this is that $\Pi_t$ *determines* the bond holdings. We define $\psi_t$, for each $t$, to be the solution of the equation

$$\Pi_t = \phi S_t + \psi P_t$$

We can consider $\Pi_T$ to be the solution of an **integral equation** constructed from equation (6.20). Hereafter, we will record this integral identity by writing it in the

### Portfolio Differential Form

$$d\Pi_t = \phi dS_t + \psi dP_t \tag{6.21}$$

This is our *definition of* $\Pi$ when we are analyzing any continuous-time investment portfolio.

# CHAPTER
# 7

# HEDGING

*I reported this to Mr. J. P. Morgan, adding that I was willing to gamble half the sum from my own funds.*

*"I never gamble," replied Mr. Morgan.*

Bernard Baruch

Every day, banks, multinational corporations, investment houses, funds, and investors enter into large financial positions. These entities and individuals wish to protect themselves from risk and uncertainty or wish to limit the risk and uncertainty to tolerable levels.

*Hedging* is a means of minimizing this risk. Hedging is a form of insurance. People hedge stocks, bonds, interest rates, commodities, and futures.

We cannot discuss the enormous range of possible hedges, but we will show how to hedge an option sale with a stock position. We begin with a very simple technique known as delta hedging.

## 7.1  DELTA HEDGING

Assume you are a dealer who has just sold a call option on 1000 shares of company ABC. You need to protect yourself against a rise in the price of ABC. You could simply buy 1000 shares. In effect, you would be writing a covered call. There are at least two things wrong with this approach:

- If the price of ABC drops, you lose money.
- Buying 1000 shares requires you to borrow a considerable sum of money.

Instead, let us consider a *delta hedge*. Here is a very simple way to look at delta hedging. Suppose you have sold a call. You observe that when the stock price goes up $1, the call price goes up $0.50, in other words, "two for one." You could balance out 100 calls with 50 shares of stock or 40 calls with 20 shares of stock.

If the call price went up $0.20 when the stock price went up $1, then this would be a "five for one" ratio. To hedge or balance 100 calls, you would only need to sell 20 shares of stock.

What is this "two for one" and "five for one" behavior in mathematical terms? It is just the ratio of the change of option price to change of stock price.

$$\text{Ratio} = \frac{\text{change in option price}}{\text{change in stock price}}$$

But then we see that

$$\text{Ratio} = \Delta = \frac{\partial C}{\partial S}$$

The idea behind the delta hedge simply entails balancing out the changes in option price and stock price so that their changes cancel.

## 7.1.1 Hedging, Dynamic Programming, and a Proof that Black-Scholes Really Works in an Idealized World

We set up a portfolio consisting of $a$ shares of stock, where we sell one option and add just enough cash or debt to make the net amount equal to zero. In other words, the value of the portfolio is

$$\Pi = aS - V + C$$

and its differential is

$$d\Pi = adS - dV + rCdt \qquad (7.1)$$

You should determine why we can replace $dC$ by $rCdt$.

We would like to choose $a$ so that $d\Pi = 0$. Recall our expression for $dV$ from Chapter 6:

$$dV = \frac{\partial V}{\partial S}dS + \frac{\partial V}{\partial t}dt + \frac{1}{2}\frac{\partial^2 V}{\partial S^2}(dS)^2 + \text{higher-order terms}$$

$$\approx \frac{\partial V}{\partial S}dS + \frac{\partial V}{\partial t}dt + \frac{\sigma^2}{2}S^2\frac{\partial^2 V}{\partial S^2}dt$$

But the Black-Scholes equation (6.10) states that

$$\frac{\partial V}{\partial t} + \frac{\sigma^2}{2}S^2\frac{\partial^2 V}{\partial S^2} = rV - rS\frac{\partial V}{\partial S}$$

so we use

$$dV = \frac{\partial V}{\partial S}dS + rVdt - rS\frac{\partial V}{\partial S}dt$$

We substitute this expression into (7.1) to obtain

$$d\Pi = adS - \left(\frac{\partial V}{\partial S}dS + rVdt - rS\frac{\partial V}{\partial S}dt\right) + rCdt$$

By choosing $a = \partial V / \partial S$, we see that

$$d\Pi = r\left(-V + Sa + C\right)dt = r\,\Pi\,dt$$

But $\Pi = 0$ initially, so $d\Pi = 0$.

We have presented another derivation of the delta hedge, but we have done much more than that. Assume that we start our hedging program at time $t = 0$ by setting up our portfolio

$$\Pi = aS - V + C$$

Now, as time changes, we instantaneously readjust the portfolio to account for the change in $V$ and $S_t$. Thus, $S$, $\Pi$, $\Delta$, and $C$ are all time dependent. By implementing these instantaneous changes in $\Delta(t)$ (to balance the changes in $S$), we ensure that $d\Pi = r\Pi dt$ at every moment in time.

But $\Pi(0) = 0$, so $\Pi(t) = 0$ for $0 \le t \le T$. To put it another way the solution to the differential equation

$$\frac{d\Pi}{dt} = r\Pi$$

with $\Pi(0) = 0$ is precisely $\Pi(t) = 0$ for all $t$. At $t = T$, our portfolio must have value 0. Thus, no matter the value of the stock, $S_T$, at time $T$, our stock and cash position must balance out against the option position.

This last remark tells us that the Black-Scholes price must be the unique correct price for the option in a world where we know $S(t)$ evolves as a geometric Brownian motion, and we can adjust $\Delta(t)$ instant by instant.

### 7.1.2   Why the Foregoing Argument Does Not Hold in the Real World

Note that the strategy we employed requires us to make an infinitesimal adjustment at every instant in time. It requires that we know $S(t_0)$ at time $t_0$ and simultaneously adjust $\Delta(t_0)$ by buying or selling stock. We do not buy or sell a discrete amount of stock. We buy or sell shares of stock at a rate

$$\frac{d\Delta(t)}{dt}$$

which is constantly changing. It clearly is not possible to trade in this fashion in the real world. Second, note that we do not pay any spread or transaction costs with our portfolio approach.

## 7.1.3 Earlier Δ Hedges

We see that in the continuous case, the desired hedge ratio is

$$\Delta = \frac{\partial V}{\partial S} \tag{7.2}$$

Looking back to Chapter 3, we see that

$$a = \frac{U - D}{S_u - S_d}$$

is the desired hedge ratio in the discrete case. But

$$U - D = \Delta V$$
$$= \text{difference in derivative price}$$

and

$$S_u - S_d = \Delta S$$
$$= \text{difference in stock price}$$

and so

$$a = \frac{\Delta V}{\Delta S} \tag{7.3}$$

Equation (7.3) is the discrete analog to (7.2).

### *Hedging Rule:*

*To hedge the sale of one option, buy Δ shares of the stock.*

Let us look at this rule in action. First, we need a way to compute $\Delta$. Fortunately, that is easy for the case of a call option. Note that we are using $\Delta = \partial V/\partial S$ from the Black-Scholes model, not the ratio $\partial V/\partial S$ based on real-world data. By carefully differentiating the Black-Scholes formula (5.10), one finds that

$$\Delta = N(d_1) \tag{7.4}$$

The derivation of this formula is a bit tricky. We provide the details in Section 7.5.

---

**Example** We sell options on 1000 shares of stock ABC where $S_0 = 50, X = 40, r = 0.05, \sigma = 0.30$, and $T = 1$ year. Thus,

$$d_1 = \frac{\ln 50/40 + (0.05 + 0.30^2/2)}{0.30} = 1.060$$

Hence,

$$\Delta = N(1.060) = 0.8554$$

Therefore, to hedge our sale of options on 1,000 shares of stock, we should purchase 855 shares of stock ABC. At this point, the call seller is protected against small changes

**TABLE 7.1**
**Delta Hedge: Stock ABC**

| Time to expiration in weeks | Stock price | $d_1$ | $\Delta = N(d_1)$ | Number of shares of stock |
|---|---|---|---|---|
| 52 | 50 | 1.060 | 0.8554 | 855 |
| 51 | 51.5 | 1.164 | 0.8778 | 878 |
| 50 | 49 | 1.00004 | 0.8413 | 841 |

in the market. However, a $\Delta$ hedge is not a "set-it-and-forget-it" hedge. It must be constantly readjusted. Let us see how that is done.

Assume we adjust our hedge on a weekly basis. The price of the stock and the calculations to obtain the hedge amount each week are shown in Table 7.1.

Let us run over what we have done. In the first week (one year until expiration), we bought 855 shares of stock. In the next week, the stock price had risen, increasing the value of each call relative to the stock. We now needed to purchase an additional 23 shares to bring the stock holdings up to 878 shares of stock.

In the following week, the stock price fell and so did $\Delta$. At this point, we need to hold only 841 shares of stock. We sell $37 = (878 - 841)$ shares of stock to bring our hedge back into line. We then continue in this fashion until expiration.

The $\Delta$ hedge has a number of drawbacks. Here are two major ones:

1. If you look carefully at Table 7.1, you can see one of the problems known as *hedge slippage*. When the stock price rises, we buy more shares of stock, and when the price drops, we sell shares of stock.

   The pattern we are locked into is *buying high and selling low*; we are always going the wrong way on stock trades. We can minimize this loss by hedging more often. But that brings us to another problem.

2. Transaction costs for hedging are substantial, since we are readjusting the position on a regular basis.

The example above illustrates a method for hedging an option sale. One option can be hedged with $\Delta$ shares of stock. On the other hand, this provides a way to hedge a stock with calls. A share of stock can be hedged with $1/\Delta$ call options. We next discuss other means of hedging a stock or a portfolio of stocks.

## 7.2   METHODS FOR HEDGING A STOCK OR PORTFOLIO

### 7.2.1   Hedging with Puts

One can simply purchase a *put* on a stock or purchase *index option puts* on a portfolio. This raises delicate questions as to what percent of the portfolio to hedge and how far out of the money to choose the put.

The cost of the hedge will vary widely depending on the selection. One could purchase puts on each stock in a portfolio, but it is much less expensive to purchase index option puts.

## 7.2.2 Hedging with Collars

To hedge a stock with a collar, one purchases an *at-the-money* put and sells an *out-of-the-money* call. Properly chosen, the cash from the sale will cover the purchase price of the put, so no further cash is required from the investor. If the stock drops, the investor is protected. If the stock increases in price, then it will be called away.

## 7.2.3 Hedging with Paired Trades

Suppose that in the same industrial group there are two companies: *Super Good* and *Hapless Industries*. It is agreed that the names accurately reflect the quality of the companies. Suppose we purchase 100 shares of Super and short an equal dollar amount of Hapless.

If the economy surges strongly, both stocks should appreciate, but Super should outperform Hapless. If the economy falters, both stocks will drop, but Hapless should fall faster and further than Super. The same analysis holds if an exogenous or endogenous shock strikes the specific industrial niche occupied by Super and Hapless. Thus, no matter what the future holds, Super should outperform Hapless on a relative basis.

At first glance, this approach seems too good to be true. We cannot lose! That is true, but if the market appreciates, we would do much better by purchasing Super alone and avoiding Hapless. So we are giving up profits to protect ourselves on the downside. This is true for all hedges.

## 7.2.4 Correlation-Based Hedges

The performance of some stocks are positively correlated, and that of others is negatively correlated. We can use this information to construct a portfolio in which the impact of the market on the portfolio is damped to the degree desired.

We might select and purchase shares of several stocks whose past price movements have been negatively correlated. The individual stock price changes would tend to cancel each other. Another approach is to sell some stocks short and buy other stocks that tend to move in the same direction as the shorted stocks.

***Observation.*** Hedging provides protection, but it compromises performance. If we have a portfolio with 200 stocks, it will probably be buffered to some extent against market moves, but it will very likely underperform more focused portfolios.

Investors such as Warren Buffett and George Soros go the other way. They make very large bets on a small number of stocks or situations. This strategy magnifies the outcome, up or down, of their investment choices. This approach is not for the fainthearted.

### 7.2.5 Hedging in the Real World

On June 29, 1998, the *Wall Street Journal* (page C1) reported on the activities of several *hedge funds*. These funds had gone long on mortgage-backed securities and had attempted to hedge the position by shorting Treasury bonds.

Since both instruments are sensitive to interest rate changes, the funds believed the hedge would protect them. However, the price of the mortgage-based securities dropped because homeowners were prepaying and refinancing their mortgages; the price of Treasury bonds rose.

Funds such as Shetland Fund, Lane Capital, Watch Ill, MKP Capital, Atlantic Portfolio Analytics, and Capstead Mortgage lost on both sides of the trade, suffering substantial losses. The sign and size of $\Delta$ in the delta hedge is intended to protect against an event of this nature.

## 7.3 IMPLIED VOLATILITY

We have already encountered one definition of volatility. Historical volatility (Section 5.3) is the standard deviation of real market prices measured over some prespecified time period.

*Implied volatility* is different. Let us assume we are given the following information for a European call option:

$$S_0 = \text{stock price today}$$
$$K = \text{strike price}$$
$$T = \text{time to expiration}$$
$$r = \text{risk-free rate}$$
$$V = \text{market price of the option today}$$

Note that we have **not** specified the volatility $\sigma$. We take $\sigma$ to be an unknown and set

$$V = S_0 N(d_1) - Ke^{-rT}N(d_2) \tag{7.5}$$

where the right-hand side is just the Black-Scholes formula but with $\sigma$ as a variable. Solve (7.5) for $\sigma$; its value is defined to be the implied volatility $\sigma_I$.

### 7.3.1 Computing $\sigma_I$ with Maple

At the initial Maple prompt, proceed as follows

```
>with(finance):
>solve (blackscholes (S_0, K, r, T, x) = V, x);
```

Maple should now print out the numerical value of $\sigma_I$.

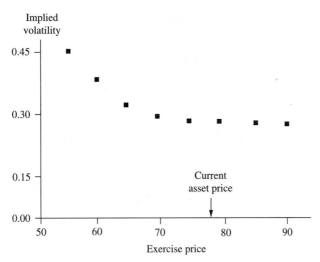

**FIGURE 7.1**
The Volatility Smile

## 7.3.2 The Volatility Smile

The Black-Scholes formula assumes that (historical) volatility is constant over the life of the option. Suppose we calculate implied volatility for a fixed stock, say, IBM. Let us look at a collection of European call options on IBM, all with the same expiration date, but with a range of strike prices. In theory, the implied volatilities for the different options should all be the same. However in practice, one tends to find the patterns displayed in Figure 7.1.

Various explanations for this phenomenon have been proposed over the years. One should bear in mind that option purchasers know much more about the market than just the historical volatility of the stock. Using specific knowledge of the firms, they may be able to forecast the future more accurately. The volatility smile may be a reflection of this extra insight and foresight.

## EXERCISE

**1.** Given the following information, find the implied volatility for the associated European call option.

|       | $S_0$ | $K$ | $r$   | $T$    | $V$  |
|-------|-------|-----|-------|--------|------|
| (*a*) | 40    | 40  | 0.05  | 3 mo   | 3.8  |
| (*b*) | 50    | 45  | 0.048 | 1 year | 9    |
| (*c*) | 80    | 85  | 0.051 | 6 mo   | 3.2  |
| (*d*) | 100   | 90  | 0.055 | 4 mo   | 13.1 |
| (*e*) | 25    | 22  | 0.052 | 1 mo   | 4.6  |
| (*f*) | 60    | 60  | 0.049 | 1 year | 3.1  |
| (*g*) | 45    | 60  | 0.047 | 3 mo   | 2    |

## 7.4   THE PARAMETERS $\Delta$, $\Gamma$, AND $\Theta$

We have just seen that $\Delta$ plays a vital role in the hedging of options. Several other quantities arising in the option context also can aid and abet the active practitioner. We will discuss two of them: $\Gamma$ (gamma) and $\Theta$ (theta).

Let us first look back at our series expansion for the call price, $C(S, t)$. It takes the form

$$dC = \frac{\partial C}{\partial t} dt + \frac{\partial C}{\partial S} dS + \frac{1}{2} \frac{\partial^2 C}{\partial S^2} (dS)^2 + \text{higher-order terms} \qquad (7.6)$$

Next, recall that the Black-Scholes equation (6.10) has the form

$$\frac{\partial C}{\partial t} dt + rS \frac{\partial C}{\partial S} dS + \frac{1}{2} \sigma^2 S^2 \frac{\partial^2 C}{\partial S^2} = rC \qquad (7.7)$$

In both equations, three expressions appear, each of which plays a very important role in the pricing and hedging of options. They are

$$\Delta = \frac{\partial C}{\partial S} = N(d_1) \qquad (7.8)$$

$$\Gamma = \frac{\partial^2 C}{\partial S^2} = \frac{1}{\sigma S \sqrt{2\pi T}} e^{-d_1^2/2} \qquad (7.9)$$

$$\Theta = \frac{\partial C}{\partial t} = -re^{-rT} XN(d_2) - \frac{1}{2}\sigma^2 S^2 \Gamma \qquad (7.10)$$

We can combine all three in one equation as

$$\Theta + rS\Delta + \frac{1}{2}\sigma^2 S^2 \Gamma = rC \qquad (7.11)$$

In these equations,

$$S = \text{stock price at } t = 0$$
$$r = \text{risk-free rate}$$
$$\sigma = \text{volatility}$$
$$T = \text{time to expiration}$$
$$X = \text{strike price}$$

We have already discussed $\Delta$ at length. How does $\Delta$ behave as the stock price varies relative to the strike price and $t \to T$ (time runs out)? The answer appears graphically in Figure 7.2.

Note that as $t \to T$, the graph of $\Delta$ becomes steeper near the strike price $X$. Also,

$$\lim_{t \to T} \Delta(t) = \begin{cases} 0 & \text{for } S < X \\ 1 & \text{for } S > X \end{cases}$$

Note that Figure 7.2 not only can be derived from the definition of $\Delta$ as $N(d_1)$ but also should be clear on financial grounds. Thus, if $S_t > X$, then the value of

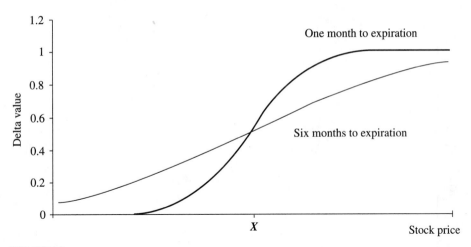

**FIGURE 7.2**
Δ versus stock price

the call approaches $S_T - X$ as $t \to T$, and the values $C$ and $S_t$ move in lockstep. Hence,

$$\Delta = \frac{\partial C}{\partial S} \approx 1$$

A similar argument holds for $S_t < X$, but then $C \approx 0$, so that $\partial C / \partial S \approx 0$.

### 7.4.1 The Role of Γ

We can use Γ either to anticipate the reaction of Δ to market changes or to implement a more nimble and nuanced hedge. To see how this might be done, consider Figure 7.3.

Note that the option price curve lies above the tangent curve to either side of the point $(S, C(S, t))$. We can see why this is true analytically by looking at equation (7.9). Γ is always positive, so the call price is a convex function of $S$.

We can express the call dependence on $S$ with the expansion

$$dC \approx \Delta dS + \frac{1}{2}\Gamma(dS)^2$$

To match the call price after sizable stock price changes more closely, we might "add some Γ." As one can see from Figure 7.4, Γ is largest when $S \approx X$ and the time to expiration is small. So we could inexpensively match the Γ of the call we are hedging by buying a *close-to-expiration, at-the-money* call option on our stock. If we do this, and readjust the stock amount to achieve the desired Δ, we now have both a Δ and a Γ hedge.

Note that as $t \to T$, the graph of Γ becomes more peaked or spiked. Also, if $S \neq X$, then

$$\lim_{t \to T} \Gamma(t) = 0$$

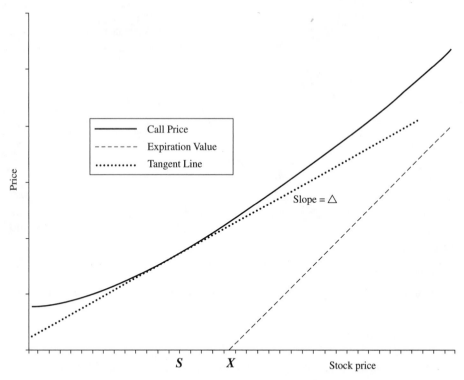

**FIGURE 7.3**
Δ as a slope

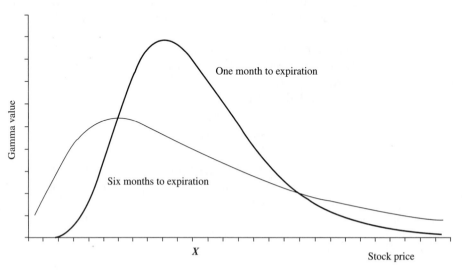

**FIGURE 7.4**
Γ versus stock price

## 7.4.2 A Further Role for Δ, Γ, Θ

Equation (7.6) gives a call price differential as

$$dC \approx \Theta dt + \Delta dS + \frac{1}{2}\Gamma(dS)^2 \qquad (7.12)$$

We can calculate a "new" option price from an older one. Let us look at an example.

| We set | We find |
|---|---|
| $S = 43$ | $C = 5.56$ (from Black-Scholes) |
| $X = 40$ | $\Delta = 0.825$ |
| $\sigma = 0.1414$ | $\Gamma = 0.143$ |
| $r = 0.05$ | $\Theta = -3.0635$ |
| $T = 1$  year | |

After three weeks, the stock price $S = 44$. Using our approximation (7.12),

$$C_{new} \simeq C_{old} + \Theta dt + \Delta dS + \frac{1}{2}\Gamma(dS)^2$$

where $dt = \frac{3}{52}$ and $dS = 44 - 43 = 1$.
  Thus,

$$C_{new} \approx 5.56 + (-3.0635)\frac{3}{52} + 0.825 \cdot 1 + \frac{1}{2}(0.143) \cdot 1^2$$

$$= 5.56 + (-0.177) + 0.825 + 0.0715$$

$$= 6.28$$

If we compute $C_{new}$ directly using Black-Scholes, we get 6.276.

## EXERCISES

**1.** Given the following values, find $\Delta$, $\Gamma$, and $\Theta$.

|  | $S_0$ | $X$ | $r$ | $T$ | $\sigma$ |
|---|---|---|---|---|---|
| (a) | 50 | 55 | 0.05 | 6 mo. | 0.30 |
| (b) | 60 | 60 | 0.052 | 3 mo. | 0.33 |
| (c) | 80 | 70 | 0.055 | 4 mo. | 0.40 |
| (d) | 40 | 50 | 0.048 | 5 mo. | 0.35 |
| (e) | 30 | 25 | 0.05 | 2 mo. | 0.25 |
| (f) | 20 | 25 | 0.053 | 5 mo. | 0.38 |

**2.** Given the following update values for the stocks in Exercise 1, find $C_{new}$ using the series approximation method in the text. Then, compare to the $C_{new}$ value computed by the Black-Scholes formula.

|     | $S_{new}$ | **Elapsed time** |
|-----|-----------|------------------|
| (a) | 51        | 2 weeks          |
| (b) | 59        | 3 weeks          |
| (c) | 82        | 4 weeks          |
| (d) | 41.5      | 6 weeks          |
| (e) | 28.5      | 2 weeks          |
| (f) | 19        | 3 weeks          |

## 7.5   DERIVATION OF THE DELTA HEDGING RULE

$$\frac{\partial V}{\partial S} = \Delta = N(d_1)$$

The verification of the relationship is not transparent. It depends on a cancellation that seems miraculous, as though it were due to some divine intervention. We begin with equation (5.10) for the call price:

$$V = SN(d_1) - e^{-rT}XN(d_2)$$

Both $d_1$ and $d_2$ are functions of $S$, so the partial derivative of $N(d)$ is the product of $N'(d)$ and $\partial d/\partial S$. In Section 6.5.1 we noted that

$$N'(d) = \frac{1}{\sqrt{2\pi}}e^{-d^2/2}$$

The product rule gives

$$\frac{\partial V}{\partial S} = N(d_1) + SN'(d_1)\frac{\partial d_1}{\partial S} - e^{-rT}XN'(d_2)\frac{\partial d_2}{\partial S}$$

Equation (5.10) contains the formulas for $d_1$ and $d_2$. Since

$$d_i = \frac{\ln S}{\sigma\sqrt{T}} + \text{constant},$$

we see that $\partial d/\partial S$ equals the factor $1/S\sigma\sqrt{T}$. Let us collect terms and see what we have:

$$\frac{\partial V}{\partial S} = N(d_1) + [SN'(d_1) - e^{-rT}XN'(d_2)]/S\sigma\sqrt{T} \tag{7.13}$$

We claim that the last two terms in equation (7.13) will cancel. This seems unlikely, but stranger things have happened. Let us look at $N(d_2)$. Equation (5.10) gives

$$d_2 = d_1 - \sigma\sqrt{T}$$

Using the formula above for $N'(d)$, we obtain

$$N'(d_2) = \frac{1}{\sqrt{2\pi}}\exp[-(d_1 - \sigma\sqrt{T})^2/2]$$

$$= \frac{1}{\sqrt{2\pi}}e^{-d_1^2/2}e^{(2d_1\sigma\sqrt{T}-\sigma^2T)/2}$$

But the formula for $d_1$ gives

$$d_1\sigma\sqrt{T} = \ln(S/X) + \sigma^2 T/2 + rT$$

We substitute this in $N'(d_2)$ to obtain

$$N'(d_2) = \frac{1}{\sqrt{2\pi}}e^{-d_1^2/2}e^{\ln(S/X)+rT}$$

$$= N'(d_1)e^{rT}S/X$$

Returning to (7.13), we find that

$$\frac{\partial V}{\partial S} = N(d_1) + [SN'(d_1) - e^{-rT}XN'(d_1)e^{rT}S/X]/S\sigma\sqrt{T}$$

$$= N(d_1)$$

## 7.6  DELTA HEDGING A STOCK PURCHASE

Let us return to our original example where we hedged a call by buying $\Delta$ shares of stock. We can *reverse* the intent of this hedge. We can hedge one share of *stock* by selling $1/\Delta$ options, where $\Delta$ is given in equation (7.8).

How expensive is it to set up this type of hedge? We have a choice of *which calls* to use in our portfolio, since the *exercise price* of a call affects its price. It seems reasonable to try to minimize the initial cost of the portfolio. Let us see whether the proper choice of an exercise price, $X$, will achieve this.

The liquidation value of the holdings at any time is given by the formula

$$H_t = \text{(Number of calls)} \times \text{(Price of a call)}.$$

We know both factors on the right-hand side of this formula. Namely,

$$\text{Number of calls} = 1/\Delta$$
$$= 1/N(d_1).$$

Using equation (5.10) for the price of a call, we obtain the value for $H_0$.

$$H_0 = [S_0 N(d_1) - e^{-rT}XN(d_2)] \times 1/N(d_1)$$

$$= S_0 - e^{-rT}X\frac{N(d_2)}{N(d_1)}$$

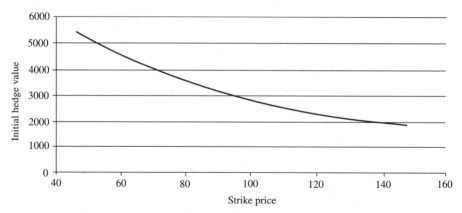

**FIGURE 7.5**
Initial hedge value versus strike price

Every term here is positive. Moreover, the ratio $N(d_2)/N(d_1)$ does not change much as $X$ varies. Thus, it appears that we minimize the value of call hedge by choosing $X$ to be large; that is, choosing *out-of-the-money* calls.

That belief is supported by Figure 7.5, which illustrates the initial investment cost when 100 shares of stock, selling at $100 per share, are hedged with calls. The graph shows that *in-the-money* calls require more initial expense. The expense falls steadily as the strike price increases. In this illustration, $\sigma = 0.38$, $r = 0.05$, and the time to expiration is one year.

To maintain this hedge over an extended period of time, we need to *reset* the liquidation amount $H_t$ periodically. For the moment, assume that adjustments are continuously made.

Since $\Delta = N(d_1)$ and $d_1$ is larger for a higher stock price, $\Delta$ *increases* as the stock price and the call price increase. But then the number of shares in our hedge, $1/\Delta$, *decreases* as the call price goes up. This means that we are on the correct side of trades each time we adjust the hedge.

> *As the call price increases, we sell calls, and as the call price decreases, we buy calls.*

It would appear that we are always making money with these trades. Actually, how much we make depends on the history of the stock price we are hedging. At any time, the total of our expense and profit from trades plus the liquidation price, $H_t$, is expected to be approximately equal to the stock price we are attempting to hedge.

# CHAPTER
# 8

## BOND MODELS AND INTEREST RATE OPTIONS

*When the sale of Thrale's brewery was going forward,
Johnson (who was an executor), on being asked what he really
considered to be the value of the property, answered: "We are
not here to sell a parcel of boilers and vats, but the potentiality
of growing rich, beyond the dreams of avarice."*

James Boswell

### 8.1 INTEREST RATES AND FORWARD RATES

Money is a commodity, and interest rates are the cost of money. Money (or capital) fuels the development of countries and industries through the construction of roads, schools, airports, hospitals, factories, telecommunication networks, power plants, steel mills, and laboratories. In general, this capital must be borrowed. This brings us to the bond market. There are many types of bonds in this country: U.S. government bonds, corporate bonds (issued by companies), and municipal bonds (issued by cities, states, hospitals, etc.) We will restrict our discussion to U.S. government bonds.

## 8.1.1   Size

The U.S. stock market is huge. The value, or **capitalization**, was about $13 trillion on January 1, 1999, and the market traded about $30 billion per day. In contrast, the U.S. government market is about $5 trillion but trades $200 billion per day. Why is the government bond market so important? It might seem, at first glance, that it was not intimately connected to the aspects of commerce mentioned above. The answer is simple. The market (or the world, if you wish) first prices government bonds, and then the rest of the bond universe is priced relative to them. If you visit the trading room at Merrill Lynch, Chase, J. P. Morgan, Citigroup, Bear Stearns, or Morgan Stanley Dean Witter, you will find the government bond desk in the middle of the floor, with the other bond groups (corporate, municipal, etc.) ringing it in satellite formation.

## 8.1.2   The Yield Curve

The yield, $Y(T)$, is simply the *annualized interest rate* one must pay today for a bond maturing in $T$ years. It is the average annualized interest rate for the period $[0, T]$. For bonds that pay no interest before they mature, the yield is computed from the ratio of the current bond price to the face (maturity) value. If $P(0, T)$ denotes this ratio, then

$$P(0, T) = e^{-T \cdot Y(T)} \qquad (8.1)$$

If we were to write interest valuations in arithmetic form, then

$$P(0, n) = [\, 1 + Y(n) \,]^{-n}$$

where $n$ is the number of years to maturity.

The yield curve for U.S. government bonds (published every day in the *Wall Street Journal*) lists the annualized rate for various time periods. We have reproduced the curve for March 11, 1999, from the *Wall Street Journal* (see Figure 8.1). By

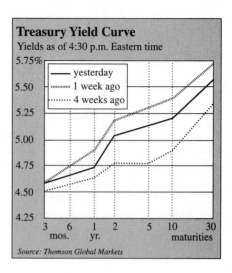

**FIGURE 8.1**
U.S. Treasury bond yield curve for March 11, 1999
(Source: *Wall Street Journal*, March 11, 1999)

examining the relative positions of the curves for one day, one week, and four weeks previously, you can see that interest rates increased from four weeks previously but fell a little in the preceding week. To be more specific, please observe that on March 10, 1999, the annualized rate for a three-month bond was about 4.67; for a one-year bond it was about 4.73, and for a 30-year bond 5.567.

A great many factors (inflation, politics, economic activity, current accounts' surplus or deficit, exchange rates) determine or influence interest rates. The shape of the yield curve is fairly normal (or not unusual); the shape can be increasing or decreasing, concave or convex, or bowl-shaped.

### 8.1.3   How Is the Yield Curve Determined?

The yield curve reflects the present interest rate or cost of money as determined by the marketplace. Dealers (or traders) on the bond desks at Morgan Stanley, Bear Stearns, Chase, and other houses are willing to buy or sell (they will go either way) for this price plus a small spread. All day long they carefully adjust prices up and down in response to changing economic conditions. The price set by traders is partly determined by supply and demand for money today, but it is also shaped by the traders' anticipations or predictions concerning the future. If they judge the market correctly, their firm profits, and they are handsomely rewarded. If they misjudge the market, their firm loses money, and "So long, it's been good to know you."

### 8.1.4   Forward Rates

Recall the notion of a forward price for a commodity such as oil. The parties agree on, say, April 1, 1999, that they will consummate a trade for 10,000 barrels of oil at a price of $15 per barrel on September 15, 1999. They agree on a *forward price* for the oil. One can also talk about a forward interest rate, although we will not be buying or selling the rate.

**Definition.** Let today be time $t = 0$. Then $f(0, t)$ denotes the interest rate at time $t > 0$ as seen from today. We refer to $f(0, t)$ as the **forward rate**.

By means of the yield curve, the bond market assigns returns or interest rates to the time period $[0, t]$. We can use the yield curve $Y(t)$ to determine the forward rate $f(0, t)$. Thus the market specifies forward rates via the yield curve. Let us carry out the details.

The expression $Y(t)$ represents the average yield for the period $[0, t]$, but the term

$$\frac{1}{t} \int_0^t f(0, s) ds$$

also represents the average yield or interest rate for this period. Thus

$$\int_0^t f(0, s) ds = t \cdot Y(t) \tag{8.2}$$

Differentiating, we find

$$f(0, t) = Y(t) + tY'(t) \tag{8.3}$$

Thus we can use the yield curve to find forward rates.

Since the yield curve is well defined only at a discrete set of points; in practice one must use some additional tools to pass from $Y$ to $f(0, t)$. We will show how this is done shortly.

## 8.2   ZERO-COUPON BONDS

Bonds in general carry coupons, but these coupons complicate the analysis, so we will look at bonds without coupons: **zero-coupon bonds** (ZCBs). A ZCB is purchased today at a certain price. At maturity (time $T$) the bond is redeemed for \$1. One obvious question arises: What is a fair price for the bond today? In a world where interest rates did not fluctuate and the short rate was $r$, then clearly

$$P_0 = e^{-rT} \tag{8.4}$$

Under these same conditions,

$$P(t) = e^{-r(T-t)} \tag{8.5}$$

using $P(t)$ to denote the price of the bond at time $t$.

Suppose we now move to a universe where we have a variable but deterministic short rate, $r(t)$. The term $r(t)$ designates the interest rate at time $t$ as seen at that time and is called the *short rate*. Then

$$P(t) = \exp\left[-\int_t^T r(s)ds\right] \tag{8.6}$$

Prices of ZCBs are readily available to investors and financial analysts. It is not hard to recover the spot rate starting with ZCB prices. We may rewrite (8.6) as

$$\int_t^T r(s)ds = -\ln P(t) \tag{8.7}$$

We differentiate with respect to $t$ to obtain

$$r(t) = \frac{d\ln P(t)}{dt}$$

Note that $r(t)$ and $P(t)$ are related in this deterministic world.

### 8.2.1   Forward Rates and ZCBs

We now move to a nondeterministic world: the real world. This next topic is extremely important. You should make a strong effort to understand the difference between spot rates and forward rates.

**FIGURE 8.2**
Time period $T$ subdivided

Let us start by combining equations (8.1) and (8.2):

$$P(0, T) = \exp[-T \cdot Y(T)]$$
$$= \exp\left[-\int_0^T f(0, s)ds\right] \tag{8.8}$$

Note that $P(0, T)$ is the price of the bond as seen from today ($t = 0$), and $f(0, s)$ is the forward rate at time $s$ as seen from today. Let us provide some intuition for the relation between rates and bond prices, as stated in equation (8.8).

We may divide the period $[0, T]$ into small blocks $[t_k, t_{k+1}]$, as in Figure 8.2, and we let $\Delta t_k = t_{k+1} - t_k$. We assume that the forward rate, $f(0, t_k)$, is roughly constant on the $k$th interval. We enter into a forward contract for each of the periods, $[t_k, t_{k+1}]$. We invest \$1 at time 0 and roll it forward at the beginning of each new period. Thus, the value, $V$, of our dollar at time $T$ is just

$$V = 1 \cdot e^{f(0, t_1)\Delta t_1} \cdot e^{f(0, t_2)\Delta t_2} \cdots e^{f(0, t_N)\Delta t_N}$$
$$= \exp\left[\sum_{k=1}^N f(0, t_k)\Delta t_k\right]$$

However, all parties should recognize the last expression (for $\Delta t_k$ small) as

$$\exp\left[\int_0^T f(0, s)ds\right]$$

Since

$$V = \exp\left[\int_0^T f(0, s)ds\right]$$

it follows that

$$P(0, T) = V^{-1} = \exp\left[-\int_0^T f(0, s)ds\right] \tag{8.9}$$

Please note that (8.9) is a real-world equation with real-world functions and real-world numbers. Any time that equation (8.9) does not hold, arbitrage will drive the bond price to this value.

If we take logarithms and differentiate with respect to $T$, it easily follows that

$$f(0, T) = -\frac{d \ln P}{dT}(0, T) \tag{8.10}$$

With no extra effort, we can generalize equation (8.9) as follows.

$$P(t, T) = \exp\left[-\int_t^T f(t, s)ds\right] \tag{8.11}$$

The term $P(t, T)$ is the price of a zero-coupon bond maturing at time $T$, as seen at time $t$. The term $f(t, s)$ is the forward rate at time $s$, as seen at time $t$. There is nothing special about time $t = 0$.

We can also derive equation (8.11) with a differential approach. We note that

$$P(t, T + \Delta T) \approx P(t, T) \cdot e^{-f(t, T_1)\Delta T}$$

where $T \leq T_1 \leq T + \Delta T$. (This follows from the mean value theorem.) Thus,

$$f(t, T_1)\Delta T = -\ln\left(\frac{P(t, T + \Delta T)}{P(t, T)}\right)$$

so that

$$f(t, T_1) = -\frac{\ln P(t, T + \Delta T) - \ln P(t, T)}{\Delta T}$$

Letting $\Delta T \to 0$, we see that

$$f(t, T) = -\frac{\partial \ln P}{\partial T}(t, T)$$

### 8.2.2 Computations Based on $Y(t)$ or $P(t)$

The determination of $f(0, t)$ from $Y$ or $P$ is, in practice, a little more challenging than we have made it appear. The functions $Y(t)$ and $P(t)$ are in fact specified for only a small number of times. Suppose we know $Y$ or $P$ at the points $t_1, t_2, \ldots, t_n$.

We cannot differentiate functions that are defined only at a discrete set of points. We could extend the definition of $Y$ or $P$ by interpolating linearly between the points $t_1, t_2, \ldots, t_n$, but that would give us a function with corners. We would like to obtain a *smoother version* of $Y$ or $P$. The standard technique used to achieve that end employs *cubic splines*. Packages in Excel, Maple®, and Mathematica® will perform the work to find the cubic spline. We illustrate that procedure below.

---

**Example** The yield curve on government bonds on some date is given in Table 8.1.

1. Find the cubic spline interpolation for $Y(t)$.
2. Using the curve $Y(t)$, find $f(0, t)$.

**Solution** (1) After starting the Maple computer application, enter

> **readlib( spline ) :**

This reads in special commands for producing a spline. Next, we enter the maturity times, and give them a name.

> **maturity : = [ .25, .5, 1, 2, 5, 10, 30 ] :**

**TABLE 8.1**
**Yield curve**

| Time | $Y(t)$ (in percent) |
|---|---|
| 3 mo | 4.6 |
| 6 mo | 4.70 |
| 1 year | 4.75 |
| 2 years | 5.0 |
| 5 years | 5.1 |
| 10 years | 5.2 |
| 30 years | 5.55 |

Now, whenever we refer to **maturity**, the program can act on this list of times. We also need the yields, and so we enter

> **percents : = [ 4.6, 4.7, 4.75, 5, 5.1, 5.2, 5.55 ] :**

To compute a curve through these points, enter

> **Yield: = spline( maturity, percents, cubic ) ;**

Notice the semicolon after this last command. It tells Maple to display the result. Here is what you see:

$$
\text{Yield} = 
\begin{cases}
4.5 + .283t + .703t^2 - .938t^3 & t < .5 \\
4.3 + 1.49t - 1.72t^2 + .676t^3 & t < 1 \\
5.12 - .972t + .748t^2 - .146t^3 & t < 2 \\
3.82 + .982t - .229t^2 - .0167t^3 & t < 5 \\
6.12 - .397t + .047t^2 - .00165t^3 & t < 10 \\
4.43 + .109t - .00357t^2 + .0000397t^3 & \text{otherwise}
\end{cases}
$$

Each polynomial is the spline function for the time interval listed to the right of the function. For example, the second row is the correct polynomial to plot when $0.5 < t < 1$. We have worked the first part of this example.

Figure 8.3 shows how the curve, "Yield," looks when you graph it.

(2) To find $f(0, t)$, the forward rate, we use equation (8.3), which states that

$$ f = Y + tY' $$

Maple will compute this expression for us if we proceed in *very small steps*. First, to obtain $Y'$, the derivative of the yields, we enter

> **Slope := diff( Yield, [t] ) ;**

We have labeled $Y'$ as *Slope*. Maple will display a group of polynomials much like those in part (1). Each line will contain the derivative of the Yield function.

To finish, we wish to multiply each $Y'$ polynomial by $t$ and add it to the Yield polynomial. Although Maple can be forced to do this, it is reluctant! We can enter

> **forward = Yield + t*Slope ;**

but the display will just restate the two separate terms. One can urge Maple to combine them by entering

> **simplify ( forward ) ;**

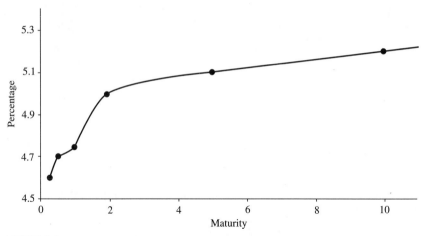

**FIGURE 8.3**
The spline obtained from the yield curve

If the numerical values do not have many decimal places, this will work. However, it is recommended that you work with each individual row, producing results such as

$$\text{yield row } 1 + t \text{ (slope row 1)}$$

to avoid hassles. Here is the final output:

$$\text{forward} = \begin{cases} 4.5 + .566t + 2.11t^2 - 3.75t^3 & t < .5 \\ 4.3 + 2.98t - 5.16t^2 + 2.71t^3 & t < 1 \\ 5.12 - 1.94t + 2.25t^2 - .584t^3 & t < 2 \\ 3.82 + 1.96t - .687t^2 + .0668t^3 & t < 5 \\ 6.12 - .794t + .141t^2 - .00165t^3 & t < 10 \\ 4.43 + .218t - .0107t^2 + .000159t^3 & \text{otherwise} \end{cases}$$

## 8.3 SWAPS

Let us begin with an example. Two companies, Slow and Steady, Inc. (SSI) and Flash Cash and Trash Cablevision (FCT), both wish to borrow $10 million. The interest rates available are displayed in Figure 8.4.

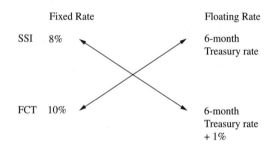

**FIGURE 8.4**
Conditions for a swap

SSI wants to borrow at a floating rate, and FCT wishes to borrow at a fixed rate. They observe, however, that

The total of FCT's fixed rate and the floating rate available to SSI is 10% + $T$

The total of FCT's floating rate and the fixed rate available to SSI is 9% + $T$

Thus, there is a difference between the ↗ and ↘ scenarios in Figure 8.4, which can possibly be exploited. Instead of SSI borrowing at the floating rate and FCT at the fixed rate, they will do the reverse: SSI will borrow at the fixed rate; FCT will borrow at the floating rate and then they will swap. More precisely, SSI will make floating-rate payments to FCT; FCT will make fixed-rate payments to SSI; and then each will resolve payments to the bank.

The only thing to be decided involves the respective payments between SSI and FCT. Before launching into that, let us look at a simpler situation.

Andy and Betty shop at different markets. They find that prices follow the rule.

|                | Apples | Oranges |
|----------------|--------|---------|
| Andy's market  | 10     | 15      |
| Betty's market | 15     | 10      |

Andy likes oranges and Betty likes apples. So Andy buys apples at his market, Betty buys oranges at her market, and they swap. As a result, both come out ahead.

This example is simple because of symmetry. Let us look at another case. Suppose the markets change their prices:

|   | Apples | Oranges |
|---|--------|---------|
| A | 10     | 15      |
| B | 11     | 13      |

It is clear that Andy should buy apples, Betty should buy oranges, and they should swap. But what should they choose as a fair price?

1. *Friendly solution:* A pays B 13c an orange; B pays A 10c an apple. A saves 2c per orange this way; B saves 1c per apple.
2. *Arithmetic average solution:* Cost for one apple and one orange = $10 + 13 = 23$. Andy pays $15 - x$ for an orange; Betty pays $11 - x$ for an apple. Thus

$$(15 - x) + (11 - x) = 26 - 2x = 23$$

So $x = 1\frac{1}{2}$.

Thus A pays $13\frac{1}{2}$ per orange (to B), B pays $9\frac{1}{2}$ per apple (to A). Both A and B save $1\frac{1}{2}$ per item this way.

3. *Geometric average solution:* A will pay $15R$ for an orange. B will pay $11R$ for an apple. Thus

$$15R + 11R = 26R = 23$$

So $R = \frac{23}{26}$.

Both get the same percent discount. So A pays $15 \cdot \frac{23}{26} = 13.27$ for an orange; B pays $11 \cdot \frac{23}{26} = 9.73$ for an apple.

Table 8.2 summarizes these cases. There could be other possible methods of allocating costs. There is no "right answer." When we turn to financial swaps, there are other complicating factors.

Let us return now to the case of SSI and FCT. How should they set up payments to one another?

Note that the difference between the ↗ and ↘ formats in Figure 8.4 is $(0.10 + T) - (0.09 + T) = 0.01 = 1\%$. Assume that SSI and FCT agree to split the 0.01 savings equally. Thus, each saves 0.005 by entering into a swap.

There is no single payment schedule for implementing this arrangement. One way to do it is shown in Figure 8.5. Since SSI "saves" 0.005 off the "standard" loan rate, we assume that SSI pays FCT the amount $(T - 0.005)$. FCT then sends SSI the amount 0.08, which is precisely what SSI needs to pay SSI's fixed loan to the bank. Thus SSI's 6-month payments net out to

$$0.08 - 0.08 + (T - 0.005) = T - 0.005$$

But where does this leave FCT? Their 6-month payments net out to

$$0.08 + (T + 0.01) - (T - 0.005) = 0.095 \qquad (8.12)$$

This again is precisely what we would have predicted. It finds FCT paying

$$0.10 - 0.005 = 0.095$$

and thus FCT "saves" 0.005 off the fixed rate.

**TABLE 8.2**
**Swap outcomes**

|   |   | Cost | | Saving | |
|---|---|---|---|---|---|
|   |   | Apple | Orange | Apple | Orange |
| 1 | A |       | 13     |       | 2      |
|   | B | 10    |        | 1     |        |
| 2 | A |       | 13 1/2 |       | 1 1/2  |
|   | B | 9 1/2 |        | 1 1/2 |        |
| 3 | A |       | 13.27  |       | 1.73   |
|   | B | 9.73  |        | 1.27  |        |

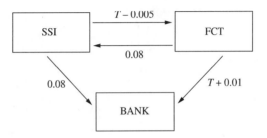

**FIGURE 8.5**
Swap transactions

## 8.3.1    Another Variation on Payments

It seems a bit inefficient to send all those checks back and forth. Let us see whether we can simplify matters: Let us have SSI send one payment to FCT to resolve matters. So we assume that SSI sends the amount $T + X$ to FCT every six months.

Then SSI's total payments equal

$$(T + X) + 0.08$$

This number should match SSI's interest charges, so

$$T + X + 0.08 = T - 0.005$$

Thus $X = -0.085$, so SSI sends FCT $(T - 0.085)$ every 6 months. Since this number is probably negative, let us say FCT sends SSI $(0.085 - T)$ every six months. Everything is okay on the SSI scene. How is FCT doing? The FCT net payment is now

$$(0.085 - T) + (T + 0.01) = 0.095.$$

So once again we see that everything matches up with one check from FCT to SSI, as shown in Figure 8.6.

Here is perhaps the simplest arrangement of all. SSI owes the bank 0.08, but we have FCT pay it for SSI. FCT in turn owes the bank $T + 0.01$, but we will have SSI pay it for FCT. To square up the balance sheet, FCT sends SSI 0.015, as shown in Figure 8.7.

You can easily check that this format finds all parties satisfied.

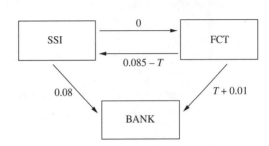

**FIGURE 8.6**
Swap with one check
between the parties

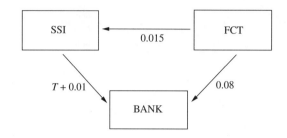

**FIGURE 8.7**
Simplest swap
arrangement

## 8.3.2 A More Realistic Scenario

Clearly SSI's credit rating is better than FCT's. By entering into a swap arrangement, SSI is taking an additional risk. FCT may default on its payments. Under the circumstances, SSI will or should expect additional compensation for the added risk. There is no free lunch. So, rather than accept a 50-50 (0.005-0.005) split of the savings on interest rates, SSI may require a (0.007-0.003) split. In this case we find

|  | Interest rate |
|---|---|
| SSI | $T - 0.007$ |
| FCT | 0.097 |

So the two companies will negotiate the swaps rate. If they cannot reach an agreement, the deal falls through. Since SSI may not wish to assume the role of bank and risk taking, it is more likely that both will buy swaps through a broker.

### EXERCISES

1. Assume that interest rates are constant for all time. In that case the price of a ZCB maturing at time $T$ in the future is

$$P(t, T) = \exp[-R(T - t)]$$

where $R$ is our constant interest rate. Verify that $f(t, T) = R$ by using equation (8.10).

2. The forward rate is given by the expression $f(0.T) = 0.05 + 0.01T$, where $T$ is measured in years. Using this information, find the price of a ZCB that matures in
   (a) 1 year
   (b) 2 years
   (c) 5 years
   (d) 10 years
   (e) 30 years

3. Assume that the forward rate curve is replaced by the equation

$$f(0, T) = 0.05 + 0.01e^{-T}$$

Repeat Exercise 2 with this new forward rate.

**4.** Andy's favorite grocery is taken over by Supersnap.com. The table below presents the new pricing structure.

|        | Apples | Oranges |
|--------|--------|---------|
| Andy   | 8      | 12      |
| Betty  | 11     | 11      |

It is agreed that Andy will buy apples, Betty will buy oranges, and they will exchange. How much should Andy and Betty charge each other under the
(a) friendly pricing method?
(b) arithmetic average pricing method?
(c) geometric pricing method?

## 8.3.3   Models for Bond Prices

Why do we need models or pricing methods for bonds? Look at the implications for swaps. In general, two companies wishing to do a swap will not team up miraculously and work out a deal. Instead, there are financial institutions, banks, and brokerages that make a business of selling swaps. A company such as SSI would negotiate a deal to swap a variable interest rate for a fixed rate or vice versa.

We now turn to the company selling the swap. Suppose they swap a variable rate for a fixed rate. To price the swap and to be successful, they must be able to "predict" or "value" future interest rates.

Even simpler, when a bank, insurance company, or private investor purchases a 30-year U.S. government bond, that entity is making a prediction or "bet" on the direction and behavior of interest rates over a 30-year period. Moreover, whatever change occurs in one part of the yield curve affects and influences the rest of the curve. To understand bond prices, one must keep this in mind.

Pricing bonds is a serious business. Mistakes are costly and swiftly punished. Therefore, bond dealers need all the help they can get.

Bond prices, yield curves, and forward rates are interchangeable, as we saw in Section 8.2. So if we can obtain a good model for one of these, we can get the others for free.

### Forward Prices for ZCB

In Section 8.2 we discussed prices, $P$, for ZCB.

Recall that $P(0, T)$ was the price today, $t = 0$, of a ZCB maturing at time $T$ and paying $1 at that time. One could define $P(t, T)$ to be the price of a ZCB at time $t$, paying $1 at time $T$. Let us carry this process one step further.

**Definition** We define $P_0(t_1, T)$ to be the price today of a ZCB issue at $t_1$, paying $1 at time $T$. Think of this as a forward contract. You agree to pay the counterparty $P_0(t_1, T)$ at time $t$, and in return, at time $t_1$ the counterparty turns the bond (piece of paper) over to you.

We could even go further and define $P_{t_1}(t_2, T)$, but let us leave that for the moment.

At first glance it might seem that it is impossible to assign a value to $P_0(t_1, T)$, because it appears to involve knowledge of the future. If so, you are in for a pleasant surprise.

From the yield curve in the newspaper (or on a hundred Web sites), we know $P(0, t)$ for $0 \leq t \leq T$.

### Proposition

$$P_0(t_1, T) = P(0, T)/P(0, t_1) \tag{8.13}$$

**Proof (Arbitrage)** We will work through the proof in careful and meticulous detail. However, we wish to emphasize that $P_0(t_1, T)$ is not a matter of discussion or intuition. It has to be $P(0, T)/P(0, t_1)$. Any deviation from that price will trigger instant arbitrage, instant pain, and ultimate profit. For brevity, we will identify a bond and its price on occasion.

**Proof in Detail** Let $\alpha = P(0, t_1)$ and $\beta = P_0(t_1, T)$. At time $t = 0$ we purchase $\beta$ units of the ZCB maturing at time $t_1$ [we buy $\beta$ units of $P(0, t_1)$]. Cost of $\beta$ units of $P(0, t_1) = \alpha\beta$.

At maturity $t_1$, we turn in the bond and receive $\beta$ dollars. At that point we pay the $\beta$ dollars and receive one bond $P_0(t_1, T)$ (we had already agreed to this trade at $t = 0$ with a counterparty). We hold our bond $P_0(t_1, T)$ to maturity and receive \$1 at time $T$. Observe that we paid out $\alpha\beta$ dollars at time $t = 0$ and received \$1 at time $T$. But we could have achieved this result by simply buying a ZCB $P(0, T)$ at $t = 0$. Thus

$$\alpha\beta = P(0, T) \tag{8.14}$$

If $\alpha\beta \neq P(0, T)$, someone can make an infinite amount of money through arbitrage as will be discussed more fully in Section 8.3.4. If we substitute in (8.14), we obtain the equation

$$P(0, t_1) \cdot P_0(t_1, T) = P(0, T)$$

Hence

$$P_0(t_1, T) = P(0, T)/P(0, t_1)$$

**Summary** The forward prices $P_0(t_1, T)$ for ZCBs are determined by either forward rates, the yield curve, or prices of ZCBs today.

## 8.3.4 Arbitrage

We remarked above that if

$$P(0, T) \neq \alpha\beta \tag{8.15}$$

the situation presents an arbitrage opportunity. Let us consider the details.

### Case 1: $\alpha\beta < P(0, T)$

In this case, buy $\beta$ units of $P(0, t_1)$ and enter into a forward contract for $P_0(t_1, T)$. At the same time, sell $P(0, T)$, pocketing the difference $P(0, T) - \alpha\beta$. Follow the course outlined in the proof. At time $T$ you receive \$1 and owe \$1, completing the trade.

**Case 2: $\alpha\beta > P(0, T)$**

In this case, sell $\beta$ units of $P(0, t_1)$ for $\alpha\beta$ and sell the forward contract $P_0(t_1, T)$. With the proceeds $(\alpha\beta)$ buy the ZCB $P(0, T)$ for $P(0, T)$ and pocket the difference, $P(0, T) - \alpha\beta$.

At time $T$, you receive $\beta$ dollars (for the bond $P_0(t_1, T)$), which you turn over to the holder of the $P(0, T)$ bond, thus completing the trade.

Continuing the two cases, we see that you, as arbitrageur, realize a profit of $[P(0, T) - \alpha\beta]$, and since you receive the amount at $t = 0$, you can invest it at $t = 0$ and be even further ahead at time $T$.

---

**Example** According to the yield curve, the one-year yield is 4.7% and the five-year yield is 5.10%. Find $P_0(1, 5)$, the forward price of this ZCB.

**Solution** From the yield curve we conclude that

$$P(0, 1) = e^{-0.047}$$

and

$$P(0, 5) = e^{-0.051 \times 5} = e^{-0.755}$$

Thus,

$$P(1, 5) = P(0, 5)/P(0, 1) = e^{-0.255} = e^{-0.755}/e^{-0.047} = 0.8122$$

---

## EXERCISES

The yield curve from the *Wall Street Journal* Credit Markets column for July 1, 1999, lists interest rates as shown in Table 8.3 and Figure 8.8.

**TABLE 8.3**
**Treasury yield curve, June 30, 1999**

| Time | Analyzed interest rate (percent) |
|------|----------------------------------|
| 3 mo | 4.75 |
| 6 mo | 5.04 |
| 1 yr | 5.07 |
| 2 yr | 5.52 |
| 5 yr | 5.66 |
| 10 yr | 5.78 |
| 30 yr | 5.97 |

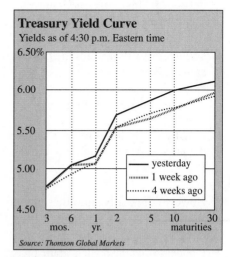

**FIGURE 8.8**
Treasury yield curve, July 1, 1999

Using the yield curve data above, find the following forward rates for ZCBs. All times are in years.

1. $P_0(\frac{1}{4}, \frac{1}{2})$. Note: $\frac{1}{4}$ yrs = 3 mo.
2. $P_0(1, 2)$
3. $P_0(1, 5)$
4. $P_0(2, 5)$
5. $P_0(5, 30)$
6. $P_0(5, 10)$
7. $P_0(1, 10)$
8. $P_0(1, 30)$
9. $P_0(1.5, 2)$
10. $P_0(10, 30)$
11. $P_0(\frac{1}{2}, 1)$
12. $P_0(2, 30)$

## 8.4   PRICING AND HEDGING A SWAP

In general, if a company wishes to purchase a swap contract it deals with a bank or investment house. Suppose a company has a loan with a floating rate. It can enter into a swap contract, which in effect exchanges the floating rate for the fixed rate. On the other side of the contract, the bank receives a fixed rate and pays a floating rate. The bank wishes to price the swap to protect itself from the unpredictable nature of future interest rates. How can this be accomplished?

The bank agrees to the following arrangement:

1. The total time period is $[0, T]$. The principal is $B_0$.
2. The bank will make payments at times $t_1, t_2, \ldots, t_N$. The periods have equal length (i.e., $t_{k+1} - t_k = \tau$), $t_0 = 0$, $t_N = T$.
3. The interest rate for the period $[t_k, t_{k+1}]$ is $R_k$, determined at $t_k$ but not known at $t = 0$.
4. The interest payment for the period $[t_k, t_{k+1}]$ is $B_0 \tau R_k$, and this payment is made at the end of the period (i.e., at $t_{k+1}$).

Two comments are in order:

1. Usually we prefer to work in the context of geometric interest rates. In that case the interest payment in item 4 in the foregoing list would be

$$B_0(e^{R_k \tau} - 1)$$

However, it greatly simplifies matters to use the arithmetic formulation for rates. We will first use arithmetic rates and then repeat the discussion with geometric rates.
2. It appears that the bank is at the mercy of changing future interest rates. However, there is a way for the bank to protect itself.

## 8.4.1 Arithmetic Interest Rates

**The Bank's Strategy**

Let us focus on the time period $[t_k, t_{k+1}]$. We do not know $R_k$ at the present time $t = 0$. The bank buys $B_0$ discount bonds $P(0, t_k)$ and sells $B_0$ discount bonds $P(0, t_{k+1})$. The cost of the position at $t = 0$ is

$$B_0[P(0, t_k) - P(0, t_{k+1})] \tag{8.16}$$

At time $t = t_k$ the bank receives \$1 for the $P(0, t_k)$ bond and buys back the $P(0, t_{k+1})$ bond, the value of which at $t_k$ is $(1 + R_k \tau)^{-1}$. Thus the net position for the bank is

$$B_0[1 - (1 + R_k \tau)^{-1}] = B_0[R_k \tau / (1 + R_k \tau)] \tag{8.17}$$

The bank invests this amount for the period $[t_k, t_{k+1}]$ at the rate $R_k$. Thus the value of the position at time $t = t_{k+1}$ is

$$B_0[R_k \tau / (1 + R_k \tau)](1 + R_k \tau) = B_0 R_k \tau \tag{8.18}$$

Note that, "miraculously," this sum is precisely the value of our floating interest payment at $t_{k+1}$. The bank implements this trade of buying $B_0 P(0, t_k)$ bonds and selling $B_0 P(0, t_{k+1})$ for every period $[t_k, t_{k+1}]$. Thus the cost at $t = 0$ is

$$\sum_{k=0}^{N-1} B_0[P(0, t_k) - P(0, t_{k+1})] = B_0[P(0, 0) - P(0, t_N)]$$

$$= B_0[1 - P(0, T)] \tag{8.19}$$

In return the bank receives a payment $B_0 r \tau$ at time $t_{k+1}$ for $k = 0, 1, \ldots, N - 1$, where $r$ is still to be determined.

These payments must be discounted, so they have value

$$B_0 r \tau P(0, t_{k+1}) \qquad \text{at } t = 0 \tag{8.20}$$

To determine $r$ we equate

$$\sum_{k=0}^{N-1} B_0 r \tau P(0, t_{k+1}) \qquad \text{and} \qquad B_0[1 - P(0, T)]$$

Thus

$$r = \frac{[1 - P(0, T)]}{\tau \sum_{k=1}^{N} P(0, t_k)} \tag{8.21}$$

Note that (8.21) is the only possible value for $r$. A different value would present opportunities for arbitrage

---

**Example** We are given the ZCB prices shown in Table 8.4. We wish to compute the interest rate $r$ we should charge on a swap contract. We will make the floating rate payments every 6 months on a loan with a fixed face amount $B_0$ for a duration of 4 years.

**TABLE 8.4**
**ZCB prices**

| Time | $P(0, T)$ (ZCB price) |
|------|------|
| 6 mo | .9977 |
| 1 year | .9953 |
| 1.5 years | .9930 |
| 2.0 years | .9906 |
| 2.5 years | .9883 |
| 3.0 years | .9860 |
| 3.5 years | .9837 |
| 4.0 years | .9814 |

**Solution** We use equation (8.21) with $\tau - 0.5$ and the values $P(0, t_k)$ from Table 8.4. Thus,

$$r = \frac{[1 - P(0, 4)]}{0.5 \sum_{k=1}^{8} P(0, t_k)}$$

$$= \frac{[1 - 0.9814]}{0.5 \cdot 7.9160} = 0.004699$$

It is easy to generalize this approach to the case where the principal and the time intervals vary with each time period. Thus we now assume that

1. The total time period is $[0, T]$. The bank will make payments at times $t_1, t_2, \ldots, t_N$. The periods have length $\tau_k = t_{k+1} - t_k$; $t_0 = 0$; and $t_N = T$.
2. The principal for the period $[t_k, t_{k+1}]$ is $B_k$.
3. As before, the interest rate for the period $[t_k, t_{k+1}]$ is $R_k$, determined at $t_k$ but not known at $t = 0$.
4. The payment for the period $[t_k, t_{k+1}]$ is $B_k \tau R_k$, and the payment is made at time $t = t_{k+1}$.

As before, at time $t = 0$, the bank buys $B_k$ of the $P(0, t_k)$ discount bonds and sells $B_k$ of the $P(0, t_{k+1})$ discount bonds for the period $[t_k, t_{k+1}]$. At time $t_k$ this position has the value

$$B_k[1 - (1 + R_k\tau_k)^{-1}] \qquad (8.22)$$

The bank reinvests this amount for the period $[t_k, t_{k+1}]$. The value of this quantity at time $t_{k+1}$ is

$$B_k[1 - (1 + R_k\tau_k)^{-1}](1 + R_k\tau_k) = B_k R_k \tau \qquad (8.23)$$

which is precisely the required floating payment. In return, the bank receives a payment $B_k r \tau_k$ at $t_{k+1}$. When discounted to time $t = 0$; this payment has value

$$B_k r \tau_k P(0, t_{k+1}) \qquad (8.24)$$

Thus

$$\sum_{k=0}^{N-1} B_k r \tau_k P(0, t_{k+1}) = \sum_{k=0}^{N-1} B_k [P(0, t_k) - P(0, t_{k+1})] \qquad (8.25)$$

This equation does not collapse as neatly as before, but let us look a little further. Note that

$$P(0, t_k) = P(0, t_{k+1})(1 + F_k \tau_k), \qquad (8.26)$$

where $F_k$ is the forward rate for the period $[t_k, t_{k+1}]$. Thus we can rewrite

$$[P(0, t_k) - P(0, t_{k+1})]$$

as

$$F_k \tau_k P(0, t_{k+1})$$

With this substitution, (8.25) becomes

$$\sum_{k=0}^{N-1} B_k r \tau_k P(0, t_{k+1}) = \sum_{k=0}^{N-1} B_k F_k \tau_k P(0, t_{k+1})$$

and hence

$$r = \frac{\sum_{k=0}^{N-1} B_k F_k \tau_k P(0, t_{k+1})}{\sum_{k=0}^{N-1} B_k \tau_k P(0, t_{k+1})} \qquad (8.27)$$

If we set

$$w_k = \frac{B_k \tau_k P(0, t_{k+1})}{\sum_{k=0}^{N-1} B_k \tau_k P(0, t_{k+1})} \qquad (8.28)$$

then we see that

$$r = \sum_{k=0}^{N-1} F_k w_k \qquad (8.29)$$

Thus $r$ is the weighted average of the forward rates.

### 8.4.2 Geometric Interest Rates

**Case 1: Constant Principal and Intervals**
As in the first arithmetic case, we assume $B_0$ is the principal for all periods and $t_{k+1} - t_k = \tau$, a constant for $k = 0, 1 \ldots, N - 1$. However, the interest payment for the period $[t_k, t_{k+1}]$, paid at $t_{k+1}$, is

$$B_0 [e^{R_k \tau} - 1] \qquad (8.30)$$

We follow the same strategy as before, buying $B_0$ of the $P(0, t_k)$ bonds and selling $B_0$ of the $P(0, t_{k+1})$ bonds at $t = 0$. At $t = t_k$ this position has the value

$$B_0 [1 - e^{-R_k \tau}] \qquad (8.31)$$

This sum is invested at the rate $R_k$ for the period $[t_k, t_{k+1}]$ as before.

At $t_{k+1}$ the value of the position is

$$B_0[1 - e^{-R_k\tau}]e^{R_k\tau} = B_0[e^{R_k\tau} - 1]$$

which is precisely our interest payment. Thus

$$\sum_{k=0}^{N-1} B_0[P(0, t_k) - P(0, t_{k+1})] = B_0[1 - P(0, \tau)]$$

$$= \sum_{k=0}^{N-1} B_0[e^{r\tau} - 1]P(0, t_{k+1})$$

Hence

$$e^{r\tau} - 1 = \frac{[1 - P(0, \tau)]}{\sum_{k=0}^{N-1} P(0, t_{k+1})} \tag{8.32}$$

and we can easily solve for $r$.

### Case 2: Variable Principal and Intervals

We introduce variable principal amounts $B_k$ and variable periods $[t_k, t_{k+1}]$, where $(t_{k+1} - t_k)$ is arbitrary.

Note that

$$P(0, t_k) = P(0, t_{k+1})e^{F_k\tau_k}$$

where $F_k$ is the forward rate for the period $[t_k, t_{k+1}]$. Arguing as before, we see that

$$e^{r\tau} - 1 = \frac{\sum_{k=0}^{N-1} B_k[e^{F_k\tau_k} - 1]P(0, t_{k+1})s}{\sum_{k=0}^{N-1} B_k P(0, t_{k+1})} \tag{8.33}$$

If we set

$$w_k = \frac{B_k P(0, t_{k+1})}{\sum_{k=0}^{N-1} B_k P(0, t_{k+1})}$$

we see that

$$e^{r\tau} - 1 = \sum_{k=0}^{N-1} w_k[e^{F_k\tau_k} - 1] \tag{8.34}$$

Thus $e^{r\tau} - 1$ is a weighted average of the forward rates $e^{F_k\tau_k} - 1$.

### EXERCISES

1. We are given the ZCB prices shown in Table 8.5. Find the interest rate that should be charged on a swap of a fixed for a floating rate. The payments must be made every 6 months on a loan with constant face value $B_0$ for a duration of 5 years.

2. Given the ZCB prices in Table 8.6, find the interest rate that should be charged on a swap of a fixed for a floating rate. The payments must be made every 3 months on a loan with constant face value $B_0$ for a duration of 3 years.

**TABLE 8.5**
**ZCB prices**

| Time | $P(0, T)$ (ZCB price) |
|---|---|
| 6 mo | 0.9971 |
| 1 year | 0.9950 |
| 1.5 years | 0.9924 |
| 2.0 years | 0.9901 |
| 2.5 years | 0.9870 |
| 3.0 years | 0.9841 |
| 3.5 years | 0.9817 |
| 4.0 years | 0.9799 |
| 4.5 years | 0.9780 |
| 5.0 years | 0.9758 |

**TABLE 8.6**
**ZCB prices**

| Time | $P(0, T)$ (ZCB price) |
|---|---|
| 3 mo | 0.9982 |
| 6 mo | 0.9969 |
| 9 mo | 0.9960 |
| 1 year | 0.9949 |
| 1.25 years | 0.9936 |
| 1.5 years | 0.9924 |
| 1.75 years | 0.9911 |
| 2.0 years | 0.9898 |
| 2.25 years | 0.9885 |
| 2.5 years | 0.9873 |
| 2.75 years | 0.9859 |
| 3.0 years | 0.9846 |

**TABLE 8.7**
**ZCB prices**

| Time | Principal (in millions) | $P(0, T)$ (ZCB price) |
|---|---|---|
| 0 | 10 | 1.0 |
| 6 mo | 9 | 0.9965 |
| 1 year | 9.5 | 0.9924 |
| 1.5 years | 10 | 0.9892 |
| 2.0 years | 8 | 0.9856 |
| 2.5 years | 9.5 | 0.9821 |
| 3.0 years | 11 | 0.9788 |
| 3.5 years | 10.5 | 0.9758 |
| 4.0 years | 0 | 0.9719 |

3. Given the ZCB prices in Table 8.7 and principal amount of the loan, find the interest rate that should be charged on a swap of a fixed for a floating rate. The payments must be made every 6 months. The principal amount of the loan is indicated in the Table. The duration of the loan is 4 years.

*Hint:* Use Equation (8.25). Solve for $r$, which is then equal to

$$\frac{\sum_{k=0}^{N-1} B_k[P(0, t_k) - P(0, t_{k+1})]}{\tau \sum_{k=0}^{N-1} B_k P(0, t_{k+1})}$$

## 8.5 INTEREST RATE MODELS

There is a wide variety of bond price models. Many of these are based on specific assumptions of how the short interest rate varies over time. These bond models based on a single interest rate all have various advantages and disadvantages, and the subject is vast and broad. We begin by looking at discrete models.

### 8.5.1   Discrete Interest Rate Models[*]

The creation of a realistic, consistent interest rate model is more challenging than it might first appear. Let's begin with a simple candidate, shown in Figure 8.9.

The first period [0, 1] has only one leg because we know the interest rate at the start of this period. Let us simplify the example by getting rid of this first leg (Figure 8.10). The numbers in the example are not very realistic, but they simplify the computation and greatly enhance the transparency of the example.

#### The Future Determines the Present

Suppose we believe that this model precisely represents the course of future interest rates. What interest rate should a bank or dealer charge for the period [0, 1] in Figure 8.10?

*Assumption.* We assume our banker or dealer is "risk indifferent." In other words, an actual dollar and an "expected dollar" have exactly the same value. Most individuals are not risk indifferent; however, "risk tolerance" varies enormously across the population.

Back to the question: Most people would automatically answer that for the period [0, 1],  $1 + r$  should equal

$$(1 + 3)/2 = 2$$

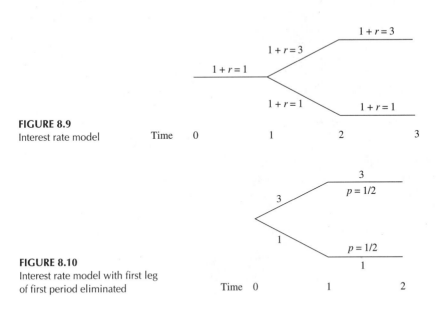

**FIGURE 8.9**
Interest rate model

**FIGURE 8.10**
Interest rate model with first leg
of first period eliminated

---

[*]Adapted in part from "Realistic Consistent Interest Models," Indiana University Technical Report, May 2000, by Joseph Stampfli.

It does not. Consider the banker who has $1 to loan. She could simply invest it in a money market fund at $t = 0$. Then

$$\text{Expected value at } t = 2 \text{ of \$1 invested at } t = 0 = \frac{1}{2}(3 \cdot 3 + 1 \cdot 1) = \$5$$

Suppose instead she loans out the $1 for the period $[0, 1]$ at the rate $1 + r = x$. At $t = 1$ she receives $x$. She invests it in a money market fund for the period $[1, 2]$. Her expected return for the period $[1, 2]$ is

$$1 + r = (3 + 1)/2 = 2$$

Thus her return on $1 for the period $[0, 2]$ with this loan and money market strategy is $2x$. To equal the pure money market strategy we described first we must have $2x = 5$, or $x = \$2.5$. Why? Recall that the future determines the present. The high interest rates are linked on the upper leg. The banker would be foolish in the extreme to accept an interest rate of less than $1 + r = 2.5$ for the period $[0, 1]$. Conversely, if the rate rises above 2.5, competition or the excess profit margin will drive it back to 2.5.

As a second example, let us modify the first example slightly, as shown in Figure 8.11. By following the previous procedure, you can and should verify that under our "risk indifferent" assumption the interest rate for the period $[0, 1]$ for this model is $1 + r = 1.5$. Note again that present interest rates are profoundly influenced by future rates.

The general discrete interest rate model appears in Figure 8.12. With this model, the interest rate for the period $[0, 1]$ must be

$$\frac{ac + bd}{c + d}$$

***Collateral.*** We have been considering the situation in which a banker loans a client $1 for the period $[0, 1]$ (or some other period). We always assume that the client provides collateral to guarantee the loan. In particular, in the United States there is a very large financial operation called the "repo" (repurchase agreement) market. This market engages in one-day or overnight loans. It works this way. A dealer loans a customer $1,000,000, and the customer in turn hands over $1,000,000

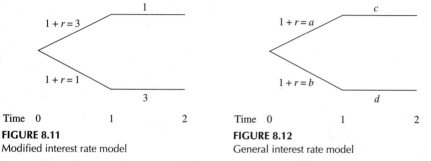

**FIGURE 8.11**
Modified interest rate model

**FIGURE 8.12**
General interest rate model

in government bonds to the dealer. The government bond belongs to the customer, and when he pays off the loan, his government bond is returned to him.

---

**Example** Consider the interest rate model in Figure 8.13. The interest rate for the period [0, 1] must be

$$\frac{ac + bd}{c + d}$$

Note that the expected value at $t = 2$ of a dollar invested at $t = 0$ is the average of the products along the branches, namely $(ac + bd)/2$.

---

**Example** In our first examples we used interest rates that had little to do with the real world. Let's look at an example with standard interest rates, illustrated in Figure 8.14.

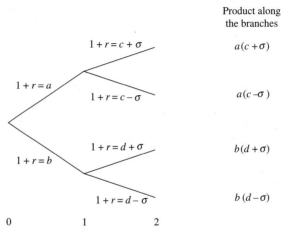

FIGURE 8.13
Product along the branches

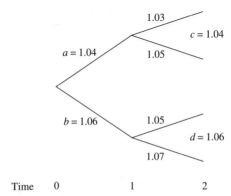

FIGURE 8.14
Model with realistic interest rates

Then the interest rate for the period [0, 1] must be

$$\frac{ac + bd}{c + d} = \frac{1.04^2 + 1.06^2}{1.04 + 1.06}$$

$$= 1.0500952$$

Note that this interest rate is only slightly larger than 1.05, the average of 1.04 and 1.06.

---

## A Specialized Interest Rate Model for Back Induction

In the next section we will want to calculate ZCB prices using back induction. To do so we must specialize the model somewhat.

***The Back Induction Model.*** In this model (Figure 8.15) the two legs emanating from any node *must* have the same value. Figure 8.14 is not an illustration of the back induction model; for the period [0, 1], $1.04 \neq 1.06$, and the other nodes fail as well.

### Interest Rates in the *N*-period Case

Suppose we are given an interest rate model (tree) with $N$ periods. How do we find the "effective," or market, interest rate $(1 + r)$ for the first period [0, 1] as we have been doing above? The answer is very simple.

**Step 1.** Find the expected value at time $t = t_N = N$ of \$1 invested at time $t = t_0 = 0$. Call this value $E[\$1 \mid t = t_N]$.

**Step 2.** Replace the two values (we called them $a$ and $b$) on the legs of the tree for the period [0, 1] with the unknown $x$. As in Step 1, compute the expected value at time $t = t_N = N$ of \$1 invested at time $t = t_0 = 0$. Call this new value $E^x[\$1 \mid t = t_N]$.

**Step 3.** Set $E^x[\$1 \mid t = t_N] = E[\$1 \mid t = t_N]$ and solve for $x$.

**Summary.** The "effective," or market, interest rate $(1 + r)$ for the period [0, 1] is simply the interest rate $(1 + r)$ for the period [0, 1] that, when

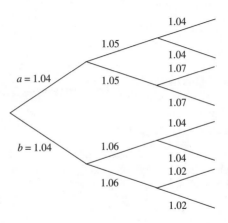

**FIGURE 8.15**
Back induction model

integrated into the original tree, produces the same expected return for the period as the original tree. Think about this summary for a minute. This is one of those situations that at first seems mysterious, but after a little thought seems obvious.

## EXERCISES

1. Verify that $1 + r = 1.5$ is the correct interest rate for the period $[0, 1]$ in Figure 8.11.
2. Verify that $1 + r = (ac + bd)/(c + d)$ is the correct interest rate for the period $[0, 1]$ in Figure 8.12.
3. Verify that $1 + r = (ac + bd)/(c + d)$ is the correct interest rate for the period $[0, 1]$ in Figure 8.13.

### 8.5.2   Pricing ZCBs from the Interest Rate Model

In this section we will present two methods (expected value and back induction) for pricing ZCBs from the interest rate model in the previous section. In Chapter 9, we will discuss the construction of interest rate models that reflect market conditions in the real world.

Before beginning on the pricing process, let's look back to Chapter 2. In that context we had (a) a stock, (b) an option, and (c) a constant interest rate environment. We were able to play one off against another to price the option. At this point all we have is an interest rate tree. It seems little to go on.

#### Method I:   Expected Value

Bear in mind that our risk-indifferent hypothesis is in force. We are given an $N$-step interest rate tree. Following the procedure just outlined, we calculate the expected value at time $t = T_N$ of \$1 invested in the money market fund at time $t = t_0 = 0$. Call this number $E[\$1 \,|[0, t_N]]$. Then the price at $t = 0$ of a ZCB paying \$1 at time $t = T_N$ is

$$1/E[\$1 \,|[0, T_N]]$$

There is an easy way to summarize and remember this process:

| Time | 0 | | $t_N$ |
|------|---|---|-------|
| If | \$1 | $\longrightarrow$ | $E[\$1 \,|[0, t_N]]$ |
| then | $\dfrac{1}{E[\$1 \,|[0, t_N]]}$ | $\longrightarrow$ | \$1 |

*Caution*: This method is valid for $t = T_N$ only when we have an $N$-step interest rate tree. For example, if $N > 1$ and you wish to calculate $P(0, 1)$, the price of the 1-year ZCB, then you must first find the "effective" interest rate $(1 + r_1)$ for the period $[0, 1]$. Then $P(0, 1) = 1/(1 + r_1)$.

---

**Example 1**   Look at Figure 8.10 in the previous section. Since \$1 $\to$ \$5 at time $t = 2$, the price of a 2-year ZCB is priced at \$0.20 at $t = 0$ by the expected value method.

---

**Example 2** Look at Figure 8.11 in the previous section. It is easy to see that $E[\$1 \,|[0, 2]]$ $= 3$. Thus the price of a ZCB at time $t = 0$ paying $1 at time $t = 2$ would be

$$\frac{1}{E[\$1\,|[0, 2]]} = \$1/3$$

## Method II: The Backward Induction Method (BIM)

The BIM applies only to backward induction interest rate models. So let us start with a simple backward induction tree, shown in Figure 8.16. Probabilities on all branches are $p = \frac{1}{2}$. The back induction process for a single "triangle" is shown in Figure 8.17. With this process,

$$P_0 = \frac{1}{2}[P_u + P_d]/a$$

In other words, we average $P_u$ and $P_d$ (since we are using probabilities $p = \frac{1}{2}$ on the branches), and then we discount by the interest rate common to the period $[k, k+1]$.

Let us go back and work out the back induction process for Figure 8.16. We began by putting in the column of 1s, which represents the value of the ZCB at $t = 2$. As shown in Figure 8.18, $P_u = \frac{1}{3}$ and $P_d = \frac{1}{4}$. Then

$$P_0 = \frac{1}{2}(1/3 + 1/4)/2$$
$$= \$7/48$$

Thus the BIM prices this 2-year ZCB at $7/48$.

It is easy to price this same bond by the expected value method. Since $E[\$1 \,|[0, 2]] = (6 + 6 + 8 + 8)/4 = 7$, the expected value method prices the bond at $1/7$.

Note that the two methods produce slightly different prices for the bond. Note that if we (a) subdivide the time period into small intervals $\Delta t$, (b) replace our interest rates $1 + r_j$ by $1 + r_j\Delta t$, and (c) let $\Delta t \rightarrow 0$, the two methods yield the same result.

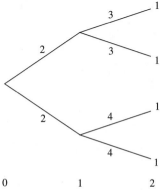

Time    0           1           2

**FIGURE 8.16**
Backward induction tree

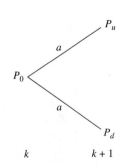

Time       $k$         $k+1$

**FIGURE 8.17**
Back induction process

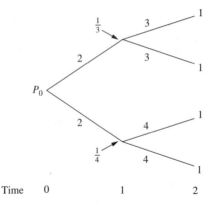

**FIGURE 8.18**
Back induction process applied to Figure 8.16

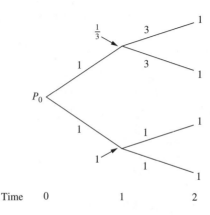

**FIGURE 8.19**
Interest rate tree

Let us look at one more example where the results are more dramatic. Consider the interest rate tree in Figure 8.19. By the BIM,

$$P_0 = \tfrac{1}{2}(\tfrac{1}{2} + 1) = \$2/3$$

By the expected value method,

$$P_0 = \$1/2$$

### The Generalized Back Induction Process

This version of the back induction method will handle an arbitrary interest rate tree. Consider the tree in Figure 8.17, but with the rate on the upper leg set to $a$ and the rate on the lower leg set to $b$. Then

$$P_0 = \tfrac{1}{2}[(P_u/a) + (P_d/b)]$$

The process then proceeds as in the original back induction method.

### 8.5.3 The Bond Price Paradox

The expected value method and the back induction method yield different values for a ZCB. Let's ask two questions: (1) Why? and (2) How would the real world resolve this issue?

*Answer 1.* In effect, the back induction method is a disguised version of Simpson's paradox. The moral of Simpson's paradox is "thou shalt not average averages." As a simple example of Simpson's paradox, consider baseball players Able and Baker. Their hitting performances are presented in Table 8.8. Note that although Able's average is higher than Baker's against both left- and right-handed pitching, Baker's average is higher against *all* pitchers. This intuition-defying switch is the crux of Simpson's paradox.

*Answer 2.* Let us return to the example in Figure 8.19. Consider a risk-indifferent investor (an investor for whom an actual dollar equals an expected dollar). She will not pay more than $0.50 for a 2-year ZCB, since she can simply invest the $0.50 in a money market fund. This investment produces an expected return of $1 in 2 years. Note that this investor does not care about this debate on ZCB pricing even if she is aware of it. Conversely, she is willing to create and sell the 2-year ZCB at any price above $0.50 (and then reinvest the proceeds in the money market fund).
 Since the risk-indifferent investor (RII) will sell at any price above $0.50 and will not pay more than $0.50 for the 2-year ZCB, the price of the bond in the real world (or, more accurately, the world of risk-indifferent investors) will be quickly driven to $0.50.

*A Plot Twist.* Perhaps this risk-indifferent investor is missing something. Consider a speculative investor (SI) who buys a 2-year ZCB from the RII for $0.60. Both are happy. At the end of 1 year the SI feels he will be ahead. After all, the expected value of the 2-year ZCB at time $t = 1$ is

$$\tfrac{1}{2}(1 + \tfrac{1}{3}) = \tfrac{2}{3} = \$0.6666$$

**TABLE 8.8**
**Able vs. Baker**

|  | Able | Baker |
| --- | --- | --- |
| Vs. left-handed pitchers: | | |
| Hits/at bats | 6/21 | 2/9 |
| Average | .286 | .222 |
| Vs. right-handed pitchers: | | |
| Hits/at bats | 5/10 | 10/21 |
| Average | .500 | .476 |
| Vs. all pitchers: | | |
| Hits/at bats | 11/31 | 12/30 |
| Average | .355 | .400 |

(Note that the money market fund value of the $0.60 at time $t = 1$ is still $0.60). But not so fast. The SI does not receive $0.666 at time $t = 1$. Let's consider the two possibilities.

*Upper Leg.*  The SI receives $0.33. What can he do with the $0.33? He can invest it at the prevailing rate $(1 + r)$ of 3.00. In this case he receives $1 at time $t = 2$. He would have also received $1 if he had held the bond so nothing has changed.

*Lower Leg.*  The SI receives $1. Again, what can he do with the $1? If he invests it at the prevailing interest rate $(1 + r)$ of 1.00, he receives $1 at time $t = 2$. But he would have received $1 at time $t = 2$ if he had held the bond. Again nothing has changed.

Our earlier comment on Simpson's paradox should be coming into focus now. The back induction method overlooks the fact that one does not receive $0.66 for the 2-year ZCB at time $t = 1$; one receives either $0.33 or $1. And one receives $1 in a "bad" interest rate environment, and receives only $0.33 in the "good" interest rate environment. It is this difference that makes the difference. Note that our risk-indifferent investor will be investing the same amount, $0.50, in both the "good" and the "bad" interest rate environments.

### 8.5.4   Can the Expected Value Pricing Method Be Arbitraged?

Consider the interest rate model in Figure 8.19. Can the price for the 2-year ZCB determined by the expected value method be arbitraged? At first glance the answer would appear to be yes. You borrow $0.50 on a 1-year bond (at the rate $1 + r = 1$) and buy a 2-year ZCB for $0.50. At time $t = 1$ your 2-year ZCB is worth $1/3 on the upper leg and $1.00 on the lower leg. Thus the expected value of the 2-year bond at time $t = 1$ is $2/3. You owe $0.50 on the 1-year bond, so it would appear you have a profit. But let us look more closely. At time $t = 1$ you must repay your obligation on the 1-year bond. If interest rates move to the upper leg at $t = 1$, your 2-year ZCB is worth only $1/3. You cannot repay the loan (represented by the 1-year bond). The person selling you the bond would have asked for collateral at time $t = 0$; without collateral he would never have sold you the bond in the first place. (However, see Exercise 6.) You can use the 2-year ZCB as partial collateral and count it as $1/3, but this means you must put up at least some collateral or money of your own.

So let us start over. You borrow $1/3, put up $1/6 of your own money, buy a 2-year ZCB for $0.50, and post this bond as collateral with the dealer who advanced the $1/3. (Incidentally, you should check that the interest rate for the period [0, 1] is $1 + r = 1$.) At time $t = 1$ you unwind the position. Let us look at the two cases confronting you.

*Upper Leg.*  The 2-year ZCB is worth $1/3. You sell it and turn the $1/3 over to the dealer to repay the loan. Your net position is now $0.

***Lower Leg.*** The 2-year ZCB is worth $1. You sell it, then turn $1/3$ over to the dealer to repay the loan. You are left with $2/3$, which you invest at the prevailing interest rate $(1 + r = 1)$. At time $t = 2$ your position is worth $2/3$. Thus, averaging over the two legs, the expected value of your initial investment of $1/6$ at time $t = 0$ is

$$\tfrac{1}{2}(0 + \tfrac{2}{3}) = \$1/3$$

So your $1/6$ has grown to $1/3$ as a result of this involved approach. But if you had simply invested the $1/6$ in the money market fund at time $t = 0$, it too would have an expected value of $1/3$ at time $t = 2$. So, at least in this situation, no arbitrage is possible.

### The General Two-Step Case

Consider the interest rate tree in Figure 8.20. The price of a 2-year ZCB at time $t = 0$ is $2/(ac + bd)$, as calculated by the expected value method. The market interest rate for the period is $(ac + bd)/(c + d)$. (You should check these assertions.) You wish to purchase a 2-year ZCB at time $t = 0$. The minimum value of this bond at time $t = 1$ is $1/c$ (since $c > d$). You will owe interest on the money you borrow; let's call the amount borrowed $x$. Thus, to make the payment at time $t = 1$, we see that

$$x\frac{ac + bd}{c + d} = \frac{1}{c}$$

so that

$$x = \frac{c + d}{c(ac + bd)}$$

Thus you must put up

$$\frac{2}{ac + bd} - x = \frac{c - d}{c(ac + bd)}$$

of your own money at time $t = 0$.

At time $t = 1$ you will sell the 2-year ZCB to pay off the loan and invest the residual at the prevailing interest rate. Let us see how that comes out.

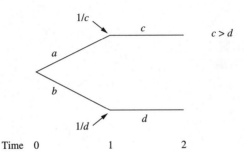

**FIGURE 8.20**
Interest rate tree          Time    0                    1                    2

***Upper Leg.*** We set things up so that the value of the 2-year bond will exactly cover the loan repayment. Your net position is 0.

***Lower Leg.*** You receive $1/d$ from the sale of the 2-year ZCB. You owe the dealer $1/c$. Thus you are left with $(1/d - 1/c)$, which you invest at the rate $d$. At time $t = 2$ your position has the value

$$d\left(\frac{1}{d} - \frac{1}{c}\right) = d\left(\frac{c - d}{cd}\right)$$

Thus, averaging over the upper and lower legs, your expected net worth is $(c - d)/2c$. Let us summarize this process:

| Initial Investment | | Final Position |
|:---:|:---:|:---:|
| $\dfrac{c - d}{c(ac + bd)}$ | $\longrightarrow$ | $\dfrac{c - d}{2c}$ |

This investment is clearly equivalent to

$$1 \longrightarrow \frac{ac + bd}{2}$$

which is precisely what you would have earned by simply investing in the money market fund. Thus, the expected value method yields an arbitrage-free price, at least in this situation.

Actually, we have shown more than it might first appear. Consider the interest rate tree in Figure 8.21. The calculations we used for Figure 8.20 show that the expected value method is arbitrage free for the more complicated tree in Figure 8.21. And the computation is exactly the same.

***Comment.*** Many authors assert that the back induction method of computing bond prices is arbitrage free. However, Musiela and Rutkowski state the model is arbitrage free only when an additional drift condition is imposed. See their

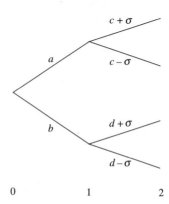

**FIGURE 8.21**
Interest rate tree            Time    0              1              2

book *Martingale Methods in Financial Modelling* (Springer, New York 1997), page 304.

## EXERCISES

**1.** (a) Given the interest rate tree, in Figure 8.22, compute the expected value at time $t = 2$ of $1 invested at time $t = 0$. Assume the probabilities assigned to all legs are $1/2$.

(b) Compute the value of a 2-year ZCB at time $t = 0$ using the expected value method.

(c) Compute the value of the 2-year ZCB in part (b) at time $t = 1$ using the expected value method.

(d) Compute the expected value at time $t = 1$ of $1 invested at time $t = 0$.

(e) Compare the results in (c) and (d).

**2.** Given the interest rate tree in Figure 8.23, price a 2-year ZCB by the back induction method. Assume the probabilities assigned to all legs of the tree are $1/2$.

**3.** For the interest rate tree in Figure 8.20, show that the interest rate for the period [0, 1] is $(a + b)/2$ precisely when $a = b$ or $c = d$.

**4.** Given the following values for $a, b, c, d, \sigma$ in Figure 8.21,

(a) Find the price of a 2-year ZCB by the expected value method.

(b) Find the "market" interest rate for the period [0, 1]

|   | a | b | c | d | $\sigma$ |
|---|-----|------|------|------|-------|
| 1 | 2 | 1 | 3 | 2 | $1/2$ |
| 2 | 1 | 2 | 3 | 2 | $1/2$ |
| 3 | 2 | 3 | 2 | 3 | 0.25 |
| 4 | 1.5 | 2 | 1.5 | 2 | 0.10 |
| 5 | 1.25 | 1.3 | 1.4 | 1.5 | 0.15 |
| 6 | 1.10 | 1.2 | 1.2 | 1.3 | 0.10 |
| 7 | 1.05 | 1.10 | 1.10 | 1.2 | 0.05 |
| 8 | 1.06 | 1.04 | 1.05 | 1.07 | 0.04 |

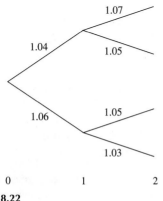

Time     0                    1                    2

**FIGURE 8.22**
Exercise 1

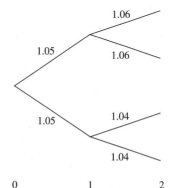

Time     0                    1                    2

**FIGURE 8.23**
Exercise 2

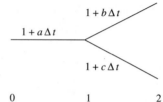

**FIGURE 8.24**
Exercise 5                              Time    0                 1                 2

5. **Expected value and backward induction for small periods.** We remarked earlier that the two methods coincide (approximately) for small time periods. Let's see why. For the tree model in Figure 8.24, the price of a 2-period ZCB is

$$P_E = \frac{1}{1 + a\Delta t} \frac{1}{1 + (b + c)\Delta t 2^{-1}} \qquad \text{expected value method}$$

and

$$P_B = \frac{1}{1 + a\Delta t}\left[\frac{1}{1 + b\Delta t} + \frac{1}{1 + c\Delta t}\right]2^{-1} \qquad \text{back induction method}$$

Since $1/(1 + a\Delta t)$ plays the same role in both expressions, let's drop it and concentrate on the remainder.

Recall that

$$\frac{1}{1 + x} = \sum_{k=0}^{\infty}(-1)^k x^k = 1 - x + x^2 - x^3 + \cdots$$

For small $x$ we can drop terms of the form $x^2$ or higher. Thus we will approximate $1/(1+x)$ by $1 - x$. Thus $P_E' \approx 1 - \Delta t(b + c)/2$. Carry out the addition inside the bracket in $P_B'$ to arrive at

$$P_B' = \frac{1 + \Delta t(b + c)/2}{(1 + b\Delta t)(1 + c\Delta t)}$$

$$\approx [1 + \Delta t(b + c)/2](1 - b\Delta t)(1 - c\Delta t)$$

$$\approx (1 + \Delta t(b + c)/2)(1 - (b + c)\Delta t)$$

and since we are dropping terms of the form $(\Delta t)^2$,

$$\approx 1 - \Delta t(b + c)/2$$

Thus $P_E'$ and $P_B'$ and hence $P_E$ and $P_B$ coincide if we ignore terms containing $(\Delta t)^k$ for $k \geq 2$.

6. Let us look at the no-arbitrage discussion of the example on page 164 (Figure 8.19). Suppose you borrow the entire $0.50 to buy the 2-year ZCB. At $t = 1$ you sell the ZCB to pay off the loan. On the upper leg you are short $1/6 ($0.50 − $1/3$). Let us assume you can borrow the $1/6 from a third party to pay off the loan. You must pay interest on this loan (at the rate $1 + r = 3$). Follow both legs through to $t = 2$. Find the expected value of your complete position at $t = 2$. You should find that you invested $0 of your own money and your expected net worth at $t = 2$ is $0. Again no arbitrage is possible.

## 8.5.5    Continuous Models

We will look at two different models out of many. In this section, we employ a fairly general approach, which we then specialize to our examples. A first simple example allows convenient calculations. Our second example is the Vasicek model.

In the next section we will change our perspective. Bond prices will become the center of attention. Using this point of view, we will then look at the very broad class of *HJM models*, introduced by David Heath, Robert Jarrow, and Andrew Morton.

As a first step, we wish to obtain a model for the price $P(t, T)$ of a zero-coupon bond. If we recall our handling of options, we will see that we can follow the same strategy.

### Review of Section 6.4

1. We started with an option with price $V(S, t)$. We assumed

$$dS = \mu S dt + \sigma S dB \qquad B = \text{Brownian motion}$$

2. We expanded $V$ in a power series, substituted $dS$, and applied Ito's lemma to discover

$$dV = \left[ \frac{\partial V}{\partial t} + \mu S \frac{\partial V}{\partial S} + \frac{1}{2}\sigma^2 S^2 \frac{\partial^2 V}{\partial S^2} \right] dt + \sigma S \frac{\partial V}{\partial S} dB$$

3. We introduced a portfolio $\Pi$ of the form

$$\Pi = V - \Delta S + C$$

or

$$\Pi = \phi S + C$$

   where $C$ denotes cash. By a judicious choice of $\Delta$, we eliminated the $dB$ term.
4. We then invoked arbitrage to end up with the differential equation

$$\frac{\partial V}{\partial t} + rS \frac{\partial V}{\partial S} + \frac{1}{2}\sigma^2 S^2 \frac{\partial^2 V}{\partial S^2} - rV = 0$$

5. We solved the differential equation.

## 8.5.6    A Bond Price Model

We will follow the same strategy for a bond price model.

### Step 1
Assume that our bond price $P(t, T)$ depends only on $T$ (maturity date), $t$, and $r(t)$, the short-term interest rate. We take as our model for $r(t)$

$$dr = \mu(r, t)dt + \sigma(r, t)dB$$

where $B$ is Brownian motion.

**Step 2**
We expand $P(t, T)$ in a power series in terms of $t, r$; substitute $dr$; and apply Ito's lemma. We find

$$dP(t, T) = \left[ \mu \frac{\partial P}{\partial r} + \frac{1}{2}\sigma^2 \frac{\partial^2 P}{\partial r^2} + \frac{\partial P}{\partial t} \right] dt + \sigma \frac{\partial P}{\partial r} dB \qquad (8.35)$$

Let us abbreviate this result by writing

$$dP(t, T) = u(t, T)\, dt + v(t, T)\, dB$$

Caution: $\mu$ and $u$ are not the same.

**Step 3. The Portfolio**
What to do? We cannot purchase units of the interest rate as a means of eliminating price uncertainty. (Actually, one can purchase $r$ as a special item, but there is no widely traded market in such instruments.) Instead, we select ZCBs with different maturities—$T_1$ and $T_2$. Thus our portfolio becomes

$$\Pi = P_1 - \Delta P_2 + C$$
$$d\Pi = dP_1 - \Delta dP_2 + rCdt$$
$$P_1 = P(t, T_1)$$
$$P_2 = P(t, T_2)$$
$$C = \text{cash}$$

Substituting the abbreviated versions of (8.35) for $dP_1$ and $dP_2$, we find that

$$d\Pi = [u_1 - \Delta u_2]dt + [v_1 - \Delta v_2]dB + rCdt$$

If we set

$$\Delta = \frac{v_1}{v_2}$$

then the $dB$ term vanishes! Thus,

$$d\Pi = [u_1 - \Delta u_2]dt + rCdt$$

Now

$$C = \Pi - [P_1 - \Delta P_2]$$

so we substitute for the cash amount. We obtain the differential expression

$$d\Pi = [u_1 - \Delta u_2]dt + r\left[ \Pi - P_1 + \frac{v_1}{v_2}P_2 \right]dt \qquad (8.36)$$

**Step 4. Elimination of Arbitrage**
If we look at (8.36), we notice that the $dB$ term has vanished, and thus $\Pi$ varies smoothly over time. Since $\Pi$ behaves like a money market investment, its return rate must match the short rate:

$$d\Pi = r(t) \cdot \Pi \, dt$$

These terms appear in equation (8.36) and cancel. We are left with the identity

$$0 = \left[ u_1 - \frac{v_1}{v_2} u_2 \right] + r \left[ -P_1 + \frac{v_1}{v_2} P_2 \right]$$

So

$$0 = u_1 - rP_1 - \frac{v_1}{v_2} [u_2 - rP_2]$$

Rearranging terms gives us

$$\frac{1}{v_1} [u_1 - rP_1] = \frac{1}{v_2} [u_2 - rP_2]$$

Now the left-hand side contains only terms involving $T_1$, and the right-hand side terms involve only $T_2$. Thus,

$$\lambda(t, r) = \frac{u(t, T) - r(t, T)P(t, T)}{v(t, T)} \tag{8.37}$$

is independent of $T$. $\lambda$ is called the *market price of risk*. We shall return to it shortly. We can rewrite the $\lambda$ term as

$$u(t, T) = rP(t, T) + \lambda v(t, T)$$

Now, use the expression for $u$ and $v$ from equation (8.35) and equate the drift terms. We arrive at the differential equation for the bond price:

$$\frac{\partial P}{\partial t} + \mu \frac{\partial P}{\partial r} + \frac{1}{2} \sigma^2 \frac{\partial^2 P}{\partial r^2} = rP + \lambda \sigma \frac{\partial P}{\partial r}.$$

We write this as

$$\frac{\partial P}{\partial t} + (\mu - \lambda \sigma) \frac{\partial P}{\partial r} + \frac{1}{2} \sigma^2 \frac{\partial^2 P}{\partial r^2} - rP = 0 \tag{8.38}$$

The terminal condition for the PDE is $P(T, T) = 1$.

Note that (8.38) looks a lot like the Black-Scholes PDE. They are both parabolic PDEs. However, (8.38) is much more general than the Black-Scholes equation, because $\mu$ and $\sigma$ are *functions* of $r$ and $t$, not constants. When we specified the initial conditions, the Black-Scholes equation had a unique solution. In contrast, there are a wide variety of solutions to (8.38), which depend on our choice for $r(t)$ as well as initial conditions.

### Step 5. Where Do We Go from Here?
We began with a model for $r$, namely

$$dr = \mu(r, t)dt + \sigma(r, t)dB$$

If we specify $\mu$, $\sigma$, and $\lambda$, we will be able to solve (8.38) in certain cases. We cannot solve explicitly for other choices of these parameters. However, we could use computational methods to obtain approximate solutions.

### 8.5.7   A Simple Example

In this example, we assume that the short rate satisfies

$$dr = \mu dt + \sigma dB \qquad (8.39)$$

where both parameters, $\mu$ and $\sigma$, are *constants*. Because these parameters are constants, we can solve equation (8.39) for $r$, and we obtain

$$r(t) = r_0 + \mu t + \sigma B_t \qquad (8.40)$$

In this case, the PDE for bond prices has constant coefficients except for two terms. These terms, from (8.38), are

$$[\mu - \lambda(t, r)\sigma]\frac{\partial P}{\partial r} \qquad \text{and} \qquad -rP$$

The PDE would be simpler to work with if $\lambda$ were constant, as well. Let us assume that the market price of risk is constant. Is this realistic? Perhaps not, but this simplification permits us to find, indirectly, an *optimal choice of constant parameters* that allows us to produce near-term price estimates.

To work with the PDE, we may as well assume that $\mu - \lambda\sigma = a$ is an unknown constant. We wish to solve the equation

$$\frac{\partial P}{\partial t} + a\frac{\partial P}{\partial r} + \frac{1}{2}\sigma^2\frac{\partial^2 P}{\partial r^2} - rP = 0 \qquad (8.41)$$

Note that (8.41) is even simpler than the Black-Scholes equation.

#### First Attempt

Let us make a guess for $P$ in equation (8.41). Suppose

$$P = e^{A \cdot r}$$

where $A$ is some constant. Could this be correct? We will see:

$$\frac{\partial P}{\partial t} = 0$$

$$\frac{\partial P}{\partial r} = Ae^{A \cdot r}$$

$$\frac{\partial^2 P}{\partial r^2} = A^2 e^{A \cdot r}$$

We substitute these in (8.41), and we obtain

$$0 + aAe^{A \cdot r} + \frac{1}{2}\sigma^2 A^2 e^{A \cdot r} - re^{A \cdot r} = 0$$

By canceling the exponential factors, we find that

$$aA + \frac{1}{2}\sigma^2 A^2 - r = 0$$

Is there a constant $A$ such that the left-hand expression is zero? No. The $r$ in this equation is a variable, while all other terms are constants. This is not a solution.

**Second Attempt**

Let us make a slightly more complicated guess for $P$. Suppose

$$P = \exp[A(t) \cdot r + B(t)]$$

and both $A(t)$ and $B(t)$ depend only on $t$. Could this be correct? For this guess, we have

$$\frac{\partial P}{\partial t} = [A'r + B']\exp[A(t) \cdot r + B(t)] = [A'r + B']P$$

$$\frac{\partial P}{\partial r} = AP$$

$$\frac{\partial^2 P}{\partial r^2} = A^2 P$$

We substitute these in (8.41) to get

$$[A'r + B']P + aAP + \frac{1}{2}\sigma^2 A^2 P - rP = 0$$

As before, we cancel the $P$'s to obtain

$$A'r + B' + aA + \frac{1}{2}\sigma^2 A^2 - r = 0$$

Now we have a mixture of constants and terms with $r$. Notice that if we take

$$A' = 1$$

then the $r$'s will cancel, and we have a solution provided that $B' + aA + \frac{1}{2}\sigma^2 A^2 = 0$. Of course, $A$ is just $t + C$ for some constant, and we make the choice

$$A(t) = t - T$$

where $T$ is the maturity date. This helps our efforts to find an answer of the form

$$P = \exp[(t - T)r + B]$$

so that $P = 1$ at maturity. We also wish to have $B(T) = 0$. But

$$B' = -aA - \frac{1}{2}\sigma^2 A^2 = -a(t - T) - \frac{1}{2}\sigma^2 (t - T)^2$$

We integrate to find that

$$B(t) = -\frac{a}{2}(t - T)^2 - \frac{\sigma^2}{6}(t - T)^3 = -\frac{a}{2}(T - t)^2 + \frac{\sigma^2}{6}(T - t)^3$$

and we have found the solution.

## Bond Prices

The preceding calculations show that

$$P(t, T) = \exp\left[-(T - t) \cdot r - \frac{a}{2}(T - t)^2 + \frac{\sigma^2}{6}(T - t)^3\right] \qquad (8.42)$$

is a solution in the special case where both $\mu$ and $\sigma$ are constant.

This simple formula *predicts* bond prices if *our input* is the interest rate. Assume that we have "correct values" for the parameters $a$ and $\sigma$. Then, at any time $t$, if we consult the market to learn the current short rate, substituting it into equation (8.42) gives "predicted" prices for bonds on that date.

Such predictions should be taken with a grain of salt for two reasons. First note that only *same-day* prices can be computed. We have only a vague notion of what the short rate will be at some future time; therefore, (8.42) offers limited help in predicting a future bond price.

Second, our model links all bond prices directly to the value of the short rate. This produces some inaccuracy in these estimates.

---

**Example** Suppose we model the U.S. Treasury market by choosing $a = 0.005$ and $\sigma = 0.03$. We are informed that $r = 0.052$ today. What prices should the 5-year and 10-year zeros be selling for today? Also, what are the current yields of these bonds?

**Solution** Since the price, in equation (8.42), depends only on the time to maturity, we use $T - t = 5$ and $T - t = 10$ in equation (8.42).
   **5-year bond:**

$$-5 \cdot (0.052) - \frac{0.005}{2}5^2 + \frac{(0.03)^2}{6}5^3 = -0.30375$$

and so $P(t, t + 5) = e^{-0.30375} = 0.738$. That is, a 5-year bond with a face value of $1,000 should be selling for $738 today. Its current yield would be $0.30375/5 = 0.0607$, a yearly return of 6.07% if held to maturity.
   **10-year bond:**

$$-10 \cdot (0.052) - (0.0025)10^2 + (0.00015)10^3 = -0.62$$

and so $P(t, t + 10) = e^{-0.62} = 0.538$. A 10-year bond with a face value of $1,000 should be selling for $538 today. Its current yield would be $0.62/10$, a yearly return of 6.2% if held to maturity.

---

## Calibration of the Simple Model

If we wish to model a specific bond market, how do we choose the parameters $a$ and $\sigma$? We observe the market yield curve on the *calibration date*, $t = 0$. We force our model to match this yield curve. This procedure has two steps.

***Step 1. The Model Yield Curve.*** Each zero-coupon bond has a current yield, $Y(t)$, that is computed from its current market price (see Section 1.4 or equation (8.1))

$$Y(t) = \frac{-\ln(P)}{T - t}$$

In our simple example,

$$\ln(P) = -(T - t)r - \frac{a}{2}(T - t)^2 + \frac{\sigma^2}{6}(T - t)^3$$

and so

$$Y(t, T) = r(t) + \frac{a}{2}(T - t) - \frac{\sigma^2}{6}(T - t)^2$$

**Step 2. Fitting the Initial Yield Curve.** The model "prediction" for the initial yield curve is

$$Y(0, T) = r_0 + \frac{a}{2}T - \frac{\sigma^2}{6}T^2$$

But we can observe actual market data on this date. For example, a Bangkok newspaper reported, on November 26, 1999, the following yields for central bank bonds:

| Term | Yield |
| --- | --- |
| 1-YR | 3.51 |
| 2-YR | 4.54 |
| 3-YR | 5.21 |
| 5-YR | 6.46 |
| 7-YR | 7.26 |
| 10-YR | 7.99 |
| 14-YR | 8.30 |

We fit the model yield curve to these data points. Figure 8.25 is a graph of the reported yields.

The "least squares" polynomial of degree 2, which best fits these points, is

$$Y = 0.02803 + 0.00892T - 0.00036T^2$$

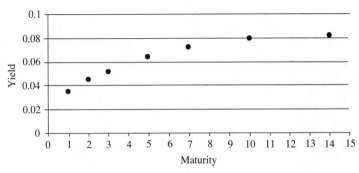

**FIGURE 8.25**
Thai bank yields

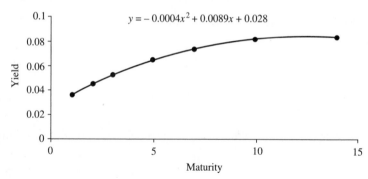

**FIGURE 8.26**
Regression curve for Thai bank yields

Its graph appears in Figure 8.26. This curve can be computed with software that handles polynomial regression. For example, Excel™ produces the results in Figure 8.10.

The coefficients determine $a$ and $\sigma$, so the bond prices become

$$P(t, T) = \exp[-(T - t)r - 0.00892(T - t)^2 + 0.00036(T - t)^3]$$

This formula, along with the short rate formula,

$$r(t) = at + \sigma B_t$$

is used to compute interest rate option prices.

### 8.5.8  The Vasicek Model

The model for interest rates in the Vasicek format is

$$dr = \alpha(\beta - r)dt + \sigma dB$$

Note that we have moved from the ultrasimple model $dr = \mu dt + \sigma dB$, with $\mu, \sigma$ constant, to a slightly more complicated model.

Such models are called "mean reverting" because the $\alpha(\beta - r)$ term pushes $r$ back to $\beta$ when it wanders away. The $\alpha$ term determines the speed or briskness of this return to $\beta$. As before, $\sigma$ reflects the volatility. To work with the Vasicek model, one assumes that the market price of risk, $\lambda(t, r)$, is a linear function of $r$. Then the nonconstant coefficient in equation (8.38),

$$\alpha(\beta - r) - \lambda\sigma$$

still has the form

$$a(b - r)$$

### The Vasicek Bond Equation
In this model, equation (8.38) has the form

$$\frac{\partial P}{\partial t} + a(b - r)\frac{\partial P}{\partial r} + \frac{1}{2}\sigma^2\frac{\partial^2 P}{\partial r^2} - rP = 0 \qquad (8.43)$$

As in the simple model case, we try to find bond prices that can be written as

$$P(t, T) = e^{A(t,T)r + B(t,T)} \qquad (8.44)$$

One finds that

$$A(t, T) = -\frac{1}{a}[1 - e^{-a(T-t)}]$$

and

$$B(t, T) = \frac{(-A(t, T) - T + t)(a^2 b - \sigma^2/2)}{a^2} - \frac{\sigma^2 A(t, T)^2}{4a}$$

Note that we have three constants, $a$, $b$, and $\sigma$. These are to be determined from market data. One obtains the yield curve at the calibration date and fits the model yield for $t = 0$ to this curve. This procedure computes $a$, $b$, and $\sigma$ implicitly. This stage of modeling is more an art than a science. You might wish to experiment and fine-tune the constants to obtain a curve for $P(t, T)$ that fits one's intuition about the market.

***Remark.*** Although we have only three parameters, $a$, $b$, and $\sigma$, at our disposal, we can still obtain a variety of interest rate curves. Figure 8.27 displays some of the possibilities.

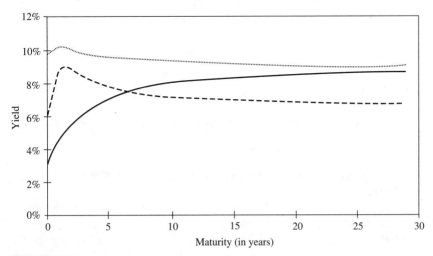

**FIGURE 8.27**
Vasicek yield curves

## EXERCISES

**1.** Check that $A(T, T) = 0$ in equation (8.44) which in turn implies that $B(T, T) = 0$. This in turn implies that $P(T, T) = 1$.

**2.** The Vasicek model can be calculated when there is no mean reversion. This is the case $a = 0$. Equation (8.44) has zero denominators in this case. But if $a$ is nearly zero, then $A(t, T) \approx t - T$. Find the approximate value for $B(t, T)$ as $a \to 0$.

## 8.6   BOND PRICE DYNAMICS

The Vasicek example illustrates that a specific short-rate model leads to bond prices. Let us reverse this process. It is advantageous first to specify a bond model and then to calculate interest rates.

How can one specify bond prices directly? One first models how bond prices change. Let us look at the *bond dynamics* of prices in the examples of Section 8.5. We will see that these resemble the dynamics of stock models.

Let us return to the five-step procedure in Section 8.5. Step 2 has a differential for the zero-coupon price $P(t, T)$:

$$dP(t, T) = u(t)dt + \sigma \frac{\partial P}{\partial r} dB$$

In step 4 of Section 8.5, we found that $u = rP + \lambda \sigma \partial P / \partial r$. Therefore,

$$dP(t, T) = \left\{ rP + \lambda \sigma \frac{\partial P}{\partial r} \right\} dt + \sigma \frac{\partial P}{\partial r} dB$$

The $\partial P / \partial r$ term now appears twice in the bond differential and is annoying, but we can eliminate it too. In both examples we found a specific formula for bond prices, of the form

$$P(t, T) = \exp[A(t, T)r + B(t, T)]$$

It is easy to find $\partial P / \partial r$.

$$\partial P / \partial r = A(t, T)\exp[A(t, T)r + B(t, T)]$$
$$= A(t, T)P(t, T)$$

Now we can write the bond differential without using any partial derivatives:

$$dP = \{r + \lambda \sigma A\}Pdt + \sigma APdB \tag{8.45}$$

Notice that $P(t, T)$ terms are the prominent feature of equation (8.45), whereas $\lambda$, $\sigma$, and $r$ have been relegated to the status of parameters.

This equation is a dynamical equation for bond prices. It expresses price changes in terms of the bond prices themselves—well, almost. The variable interest rate, $r$, and the market price of risk, $\lambda$, still appear. Nevertheless, the reader should compare equation (8.45) with equation (5.1), the GBM stock model. Despite some similarities, the bond equation is more complicated.

- *The volatility coefficient:* In contrast to the GBM stock model, the bond's volatility is not constant. The volatility, $\sigma A(t, T)$, depends on $t$. Recall from equation (8.44) that

$$\sigma A(t, T) = -\frac{\sigma}{a}(1 - e^{-a(T-t)})$$

in the Vasicek model. We see that at $t = 0$ the bond volatility is largest, while at maturity it vanishes.

      Time-dependent volatility is a realistic feature of bond models. As maturity nears, the price approaches its face value. Before maturity, bond trading produces significant volatility. Bonds behave like stocks in advance of maturity.

- *The drift coefficient:* Unlike the GBM stock model, the bond's drift is not constant. The principal reason is the behavior of the short rate—it varies over time.

      There is a model choice that simplifies the drift. Such a "calibration choice" was also carried for the GBM stock model in Chapter 5.

### Simplifying the Bond Drift

As we mentioned in examples of the previous section, there are parameters $\lambda$, $a$, $b$, and $\sigma$, which must be chosen to calibrate the model. In each case, we assumed that the market price of risk, $\lambda(t)$, could be set to 0 by adjusting another parameter value.

      Making this choice in equation (8.45), we can express the dynamics in a simple and yet general form:

### Bond Dynamics

$$dP(t, T) = r(t)P(t, T)dt + \sigma(t, T)P(t, T)dB(t) \qquad (8.46)$$

We abbreviate the volatility function, $\sigma A(t, T)$, by, $\sigma(t, T)$. This volatility function is the most important parameter that specifies a bond model. Choices other than the Vasicek formula for $\sigma(t, T)$ are often used; such models can be quite complicated.

      What is the significance of equation (8.46)? It is a differential equation for the bond price. One can usually solve a differential equation and adjust the solution to hit any target initial condition.

      In the case of bonds, we have seen the natural initial condition. We are informed of the yield curve at time $t = 0$, so we know the actual zero-coupon bond prices then. These prices make up the initial term structure.

      Equation (8.46) permits us to find models that exactly coincide with observed bond prices of all maturities *on the calibration date*. This does not happen when one uses the short rate approach to bond models. For example, with the Vasicek model one obtains only an approximate fit of the yield curve.

## 8.7 A BOND PRICE FORMULA

Can one exactly fit a bond model to the current term structure? A promising approach is to begin with the dynamical equation (8.46) and attempt to solve it. Although this equation contains an unknown drift coefficient, $r(t)$, we can ignore it at first and solve for it later.

### Step 1. The Differential of ln *P*

Equation (8.46) will give us the dynamics of $\ln P(t, T)$ as $t$ varies. We know that

$$\frac{\partial \ln P}{\partial P} = \frac{1}{P} \quad \text{and} \quad \frac{\partial^2 \ln P}{\partial P^2} = \frac{-1}{P^2}$$

We then expand $\ln P$ using first- and second-order terms:

$$d \ln P \approx \frac{1}{P} dP - \frac{1}{2P^2} (dP)^2$$

Equation (8.46) tells us what the $dP$ factor is:

$$dP = rP dt + \sigma P dB \quad \text{and} \quad (dP)^2 = \sigma^2 P^2 dt \quad \text{(by Ito's lemma)}$$

Neat cancellation occurs when these terms are combined:

$$d(\ln P) = \{r - \sigma^2/2\} dt + \sigma dB \tag{8.47}$$

### Step 2. Eliminating *r* to find ln *P*

Now, let us solve equation (8.47) for $\ln P$. Since $r(t)$ is unknown, our solution will not be useful, or so it seems. But by integrating term by term in (8.47), one finds that

$$\ln P(t, T) - \ln P(0, T) = \int_0^t r(s) ds - \frac{1}{2} \int_0^t \sigma^2(s, T) ds + \int_0^t \sigma(s, T) dB(s) \tag{8.48}$$

Any choice of $T$ is valid in equation (8.48), but there is a bonus when we use $T = t$. We know that $P(t, t) = 1$, so the LHS of (8.48) is a known constant in this case.

This allows us to obtain a formula for the annoying $r(s) ds$ term:

$$\int_0^t r(s) ds = -\ln P(0, t) + \frac{1}{2} \int_0^t \sigma^2(s, t) ds - \int_0^t \sigma(s, t) dB(s)$$

When we substitute this into equation (8.48), the annoying term disappears:

$$\ln P(t, T) = \ln \frac{P(0, T)}{P(0, t)} + \frac{1}{2} \int_0^t \{\sigma^2(s, t) - \sigma^2(s, T)\} ds + \int_0^t \{\sigma(s, T) - \sigma(s, t)\} dB(s)$$

### Step 3. The Bond Price

The equation above is a formula for $\ln P$, so we have an exponential formula for bond prices:

**Bond Price**

$$P(t, T) = \frac{P(0, T)}{P(0, t)} \exp \left[ \frac{1}{2} \int_0^t \{\sigma^2(s, t) - \sigma^2(s, T)\} ds + \int_0^t \{\sigma(s, T) - \sigma(s, t)\} dB(s) \right]$$

$$\tag{8.49}$$

The reader should ask, "How useful is this answer?" Let us think of equation (8.49) as composed of three factors. The first is

$$\frac{P(0, T)}{P(0, t)}$$

This factor is benign; we know the bond prices on the calibration date when we set up the bond model. The second factor is

$$\exp\left[\frac{1}{2}\int_0^t \{\sigma^2(s, t) - \sigma^2(s, T)\}ds\right]$$

We made a simple choice for $\sigma(s, T)$, so the integral reduces to a formula. What guides our choice for the function $\sigma(s, T)$? This is the bond trading volatility. We should use trading data to select it.

The third factor is

$$\exp\left[\int_0^t \{\sigma(s, T) - \sigma(s, t)\}dB(s)\right]$$

This factor is always random and seems to present a problem. How can a random price be useful? Here are two ways:

- Other prices are random and the bond might depend directly on these other prices. For instance, the short rate should be thought of as random. In the Vasicek example, the bond price is linked to the short rate. If you know one, you can calculate the other.
- We can use a random factor and a chaining argument to find additional prices. For instance, a call on a bond before it matures is a bond derivative. Equation (8.49) could be used to calculate the call price.

## 8.8 BOND PRICES, SPOT RATES, AND HJM

In the previous two sections, the short rate $r(t)$ was ignored even though it appeared in basic equations. But there is a compatible short rate once we have set up the bond model. Since

$$P(t, t + \Delta t) \approx e^{-r(t)\Delta t}$$

we can calculate the short rate by differentiating the exponent of equation (8.49).

### Short Rate

$$r(t) = -\frac{\partial \ln P}{\partial T}(t, T)\Big|_{T=t} \tag{8.50}$$

We will illustrate this relation shortly with an example. Note that we have achieved a goal announced in Section 8.6. It is possible to find bond prices first and then find a corresponding short rate process. This is an essential feature of the Heath, Jarrow, and Morton approach to term structure modeling.

In the HJM approach, equation (8.49) is set up using a choice for the volatility of forward interest rates. A compatible short rate is calculated in terms of these volatilities as well.

### 8.8.1   Example: The Hull-White Model

We illustrate the relation between $P(t, T)$ and $r(t)$ for the following volatility choice

$$\sigma(t, T) = \frac{\sigma}{a}[1 - e^{-a(T-t)}]$$

This bond volatility appeared in the Vasicek example. However, both $P(t, T)$ and $r(t)$ will be different from the Vasicek formulas, since we are fitting a model *exactly to a given initial term structure.*

**Bond Prices**

To begin, we calculate the three terms for $\ln P$ in equation (8.49). The crucial integrals are

$$\int_0^t \sigma^2(s, T)ds = \frac{\sigma^2}{a^2}[t + D(t, T) - D(0, T)] + \frac{\sigma^2}{2a}[D(t, T)^2 - D(0, T)^2]$$

and

$$\int_0^t \sigma(s, T)dB = \frac{\sigma}{a}\left[B(t) - e^{-aT}\int_0^t e^{as}dB\right]$$

In the first answer, the formula $D$ is the term

$$D(t, T) = \frac{1}{a}(1 - e^{-a(T-t)}) \tag{8.51}$$

This answer is basically a sum of specific exponentials. In fact, let us abbreviate the first answer by $C(t, T)$. So,

$$C(t, T) = \int_0^t \sigma^2(s, T)ds \tag{8.52}$$

is a sum of exponentials. According to (8.49),

$$\ln P = \ln P(0, T) - \ln P(0, t) + \frac{1}{2}\{C(t, t) - C(t, T)\}$$

$$+ \frac{\sigma}{a}\left[e^{-at}\int_0^t e^{as}dB - e^{-aT}\int_0^t e^{as}dB\right]$$

$$= \ln P(0, T) - \ln P(0, t) + 1/2\{C(t, t) - C(t, T)\} \tag{8.53}$$

$$+ \frac{\sigma}{a}[e^{-at} - e^{-aT}]\int_0^t e^{as}dB$$

Notice that the only nonexplicit term in these bond prices is the (random) expression

$$\int_0^t e^{as}dB(s)$$

The $\ln P(0, \cdot)$ terms are known constants since they are the observed market prices used in calibrating the model. We will see that the random term also appears in the expression for $r(t)$. For now, we can write out the bond price using equation (8.53):

$$P(t, T) = \frac{P(0, T)}{P(0, t)} \exp\left[ C(t, t)/2 - C(t, T)/2 + \sigma D(t, T)e^{-at} \int_0^t e^{as} dB \right] \quad (8.54)$$

Both $C(t, T)$ and $D(t, T)$ are explicit formulas that we could compute from equations (8.51) and (8.52).

**Short Rate**

Equation (8.50) tells us to differentiate the *negative log* of (8.54) to obtain $r$. Since the derivative is with respect to $T$, terms containing only $t$ drop out. We have

$$\frac{\partial \ln P}{\partial T} = \frac{\partial \ln P}{\partial T}(0, T) - \frac{1}{2}\frac{\partial C}{\partial T}(t, T) + \sigma e^{-aT} \int_0^t e^{as} dB$$

We find $r$ by setting $T = t$, and we obtain the *semi*-explicit expression

$$r(t) = -\frac{\partial \ln P}{\partial T}(0, t) + \frac{\sigma^2}{4a^2}D^2(0, t) - \sigma e^{-at}\int_0^t e^{as} dB \quad (8.55)$$

As in the case of the bond price, the only random term for $r$ is our familiar

$$\int_0^t e^{as} dB(s)$$

But this random term appears in two equations, (8.54) and (8.55), and so we can substitute $r$ for it.

First, we rearrange terms in (8.55) to obtain

$$\sigma e^{-at}\int_0^t e^{as} dB = -\frac{\partial \ln P}{\partial T}(0, t) + \frac{\sigma^2}{4a^2}D^2(0, t) - r(t)$$

Using this for the random term in the bond price gives us a tidy formula:

$$P(t, T) = \frac{P(0, T)}{P(0, t)}\exp\left[ C(t, t)/2 - C(t, T)/2 \right]$$
$$\cdot \exp\left[ -r(t)D(t, T) - D(t, T)\frac{\partial \ln P}{\partial T}(0, t) + \frac{\sigma^2}{4a^2}D^2(0, t)D(t, T) \right] \quad (8.56)$$

As we mentioned before, $C(t, T)$ and $D(t, T)$ are simple formulas that appear in equations (8.51) and (8.52).

Equation (8.56) contains three terms that one observes on *the calibration date*. Both expressions $P(0, T)$ and $P(0, t)$ are market bond prices at time 0. These correspond to two different maturities related to the yield curve on this date. Even the expression

$$-\frac{\partial \ln P(0, t)}{\partial T}$$

in (8.56) is simply the forward rate *observed at the moment of calibration*. The reader should look at equation (8.10) to review the relation between forward rates and bond prices.

Each bond price has only one random term: the short rate. Its "market value" determines each bond price; the other coefficients are specific functions of $t$. The Hull-White model is sometimes referred to as the *extended Vasicek model* because bond prices in both models have the same volatility.

Once we calibrate this model, only the current value of the short rate, $r(t)$, is needed for predicting bond prices at any time $t$.

### HJM Caveat

The bond formula, equation (8.49), is valid for many choices of $\sigma(t, T)$. The example above uses a nice choice, and the bond depends directly on $r(t)$. However, other choices of $\sigma(t, T)$ produce more complicated bond prices. In some HJM models the bond price depends on a history of prices, as opposed to a current price.

The exercises below make the assumption that bond volatility is deterministic. In this case, we can easily compute bond prices. However, there are choices where a bond price depends *on the entire history of the short rate*, for example, $\sigma = \sqrt{T - t}$.

### EXERCISES

1. Suppose that a bond model uses the bond volatility $\sigma(t, T) = T - t$. Use equation (8.49) to show that

$$r(t) = -\frac{\partial \ln P}{\partial T}(0, t) + \frac{t^2}{2} - B(t)$$

2. Show that the bond model in Exercise 1 has the following yield curve at time $t$:

$$Y(t) = \frac{-1}{T - t} \ln \frac{P(0, T)}{P(0, t)} - \frac{tT}{2} - B(t)$$

   **Hint:** $Y = -\ln P / (T - t)$.

3. Suppose a bond model has a bond volatility $\sigma(t, T) = t(T - t)$. Use equation (8.49) to show that

$$r(t) = -\frac{\partial \ln P}{\partial T}(0, t) + \frac{t^4}{12} - \int_0^t s\, dB(s)$$

4. Suppose a bond model has a bond volatility $\sigma(t, T) = \sqrt{T - t}$. Verify the claim made above that the bond price is not merely a function of the short rate.

5. Solve equation (8.46) for $P$ when $\sigma(t, T) \equiv 0$.

## 8.9   THE DERIVATIVE APPROACH TO HJM: THE HJM MIRACLE

We start with the bond price dynamics, according to equation (8.46):

$$dP(t, T) = r(t)P(t, T)dt + \sigma(t, T)P(t, T)dB$$

In shorthand, $dP = rP dt + \sigma P dB$.

This model copies the stock price model, except that $r$ is now a function of $t$ and $\sigma$ depends on both $t$ and $T$.

As before, we expand $\ln P$ in a power series. Our series for a general function $G$ in this context would be

$$dG = \frac{\partial G}{\partial P}dP + \frac{1}{2}\frac{\partial^2 G}{\partial P^2}(dP)^2 + \cdots$$

Thus

$$
\begin{aligned}
d \ln P &\approx \frac{1}{P}dP + \left(-\frac{1}{2}\right)\frac{1}{P^2}(dP)^2 \\
&\approx \frac{1}{P}(rPdt + \sigma PdB) - \frac{1}{2}\frac{1}{P^2}(rPdt + \sigma P\, dB)^2
\end{aligned}
\tag{8.57}
$$

With a little algebra and an appeal to Ito's lemma, this expression becomes

$$d \ln P = [r - \tfrac{1}{2}\sigma^2]dt + \sigma\, dB \tag{8.58}$$

We present two derivations of the next argument. Here is the quick one. Recall that

$$P(t, T) = \exp\left[-\int_t^T f(t, s)\, ds\right]$$

and hence

$$f(t, T) = -\frac{\partial}{\partial T}\ln P(t, T)$$

Hence

$$
\begin{aligned}
df(t, T) = -d\left(\frac{\partial}{\partial T}\ln P\right) &= -\frac{\partial}{\partial T}(d \ln P) \\
&= -\frac{\partial}{\partial T}\left\{\left[r(t) - \frac{1}{2}\sigma^2(t, T)\right]dt + \sigma(t, T)\, dB\right\} \\
&= \sigma\frac{\partial\sigma}{\partial T}dt - \frac{\partial\sigma}{\partial T}dB
\end{aligned}
$$

Replacing $B$ by $-B$, since $B$ is random and symmetric, we reach

$$df(t, T) = \sigma\frac{\partial\sigma}{\partial T}dt + \frac{\partial\sigma}{\partial T}dB \tag{8.59}$$

We present a more leisurely derivation of (8.59) in the following section.

Let us pause and review. Look at (8.59). It is a model for forward rates, but we did not assume (8.59) as an axiom; we derived it from the other models (and their properties). However, we are not done.

Let us rewrite (8.59) for the moment as

$$df(t, T) = \mu(t, T)dt + \nu(t, T)\, dB \tag{8.60}$$

One might expect $\mu$ and $\nu$ to be unrelated, but a surprise lurks just over the horizon. Let us apply the fundamental theorem of calculus to $\partial\sigma(t, T)/\partial T$. Thus

$$\sigma(t, T) - \sigma(t, t) = \int_t^T \frac{\partial\sigma(t, s)}{\partial s} ds$$

But $\sigma(t, t) = 0$, since the volatility goes to zero as the bond approaches maturity. So

$$\sigma(t, T) = \int_t^T \frac{\partial\sigma(t, s)}{\partial s} ds \qquad (8.61)$$

Let us combine (8.59), (8.60), and (8.61). We obtain

$$\mu(t, T) = \sigma(t, T)\frac{\partial\sigma(t, T)}{\partial T} = \frac{\partial\sigma(t, T)}{\partial T} \cdot \int_t^T \frac{\partial\sigma(t, s)}{\partial s} ds = \nu(t, T) \cdot \int_t^T \nu(s) ds$$

Suppressing all the variables,

$$\mu = \nu \int \nu$$

This is the HJM miracle. In this model for forward rates, the volatility determines the drift.

Does this result have any practical importance? Yes; it says one can model bond prices, forward rates, etc., by just measuring volatility and then matching short-term rates. We will look at this application in the next section. For the moment, let us take another look at (8.59):

$$df(t, T) = \sigma\frac{\partial\sigma}{\partial T}dt + \frac{\partial\sigma}{\partial T}dB \qquad (8.62)$$

It still contains the random (Brownian) term $dB$, so $f$ is not completely determined by $\sigma(t, T)$. This is unfortunate, since we know that

$$P(t, T) = \exp\left[-\int_t^T f(s, T)\, ds\right]$$

and we were hoping to find a formula for $P$. But this is only part of the problem. To proceed further, we must obtain an expression for $\sigma(t, T)$. At present, this usually means numerical methods, which leave much to be desired.

## 8.10   APPENDIX: FORWARD RATE DRIFT

We will derive equation (8.59) using bond prices directly.

We introduce the forward rate $F(t, T_1, T_2)$, which, by definition, equals the (forward) interest rate for the period $[T_1, T_2]$ as seen from $t$. Note that

$$F(t, T_1, T_2) = \frac{\ln P(t, T_1) - \ln P(t, T_2)}{T_2 - T_1} \qquad (8.63)$$

since

$$P(t, T_1) = P(t, T_2) \cdot \exp[-(T_2 - T_1)F(t, T_1, T_2)]$$

It follows immediately from (8.58) that

$$d \ln P(t, T_1) = [r(t) - \tfrac{1}{2}\sigma^2(t, T_1)]dt + \sigma(t, T_1)dB \qquad (8.64)$$

and

$$d \ln P(t, T_2) = [r(t) - \tfrac{1}{2}\sigma^2(t, T_2)]dt + \sigma(t, T_2)dB$$

Combining the terms in (8.64), we find that

$$dF(t, T_1, T_2) = \frac{\sigma^2(t, T_2) - \sigma^2(t, T_1)}{2(T_2 - T_1)}dt + \frac{\sigma(t, T_1) - \sigma(t, T_2)}{(T_2 - T_1)}dB \qquad (8.65)$$

You should carry out the algebra. Note that the $r(t)$'s have canceled; this is a bit of serendipity and quite unexpected. We take the limit as $T_2$ approaches $T_1$ to obtain

$$dF(t, T_1) = \sigma(t, T_1)\frac{\partial \sigma(t, T_1)}{\partial T_1}dt - \frac{\partial \sigma}{\partial T_1}(t, T_1)dB$$

since $f(t, T_1) = F(t, T_1, T_1)$. Now we change $T_1$ to $T$ and change the sign on $dB$ from minus to plus (since it is random and symmetric). Thus

$$df(t, T) = \sigma\frac{\partial \sigma}{\partial T}dt + \frac{\partial \sigma}{\partial T}dB \qquad (8.66)$$

# CHAPTER
# 9

## COMPUTATIONAL
## METHODS
## FOR BONDS

*Stocks have reached what looks like a permanently high plateau.*

Irving Fisher, October 17, 1929

## 9.1   TREE MODELS FOR BOND PRICES

Differential equations can be set up for bond models, but it is not possible, in most cases, to solve these differential equations to obtain a formula or closed-form solution. Consequently, analysts have turned to numerical methods such as tree or lattice expansions to obtain approximate solutions. These methods can appear imposing at first glance, so we will begin with a gentle introduction. Indeed, our first foray seems to have nothing to do with bond pricing.

### 9.1.1   Fair and Unfair Games

A fair game is a two-person game in which the average net winnings of each player is zero. In other words, neither player has an advantage.

**190**

**Example 1** Perhaps the best-known fair game is coin tossing with a fair coin. I win $1 if the toss is a head; you win $1 if the toss is a tail.

| I | II | III | Product |
|---|---|---|---|
| Outcome | Probability | Payoff | II × III |
| H | $1/2$ | +1 | +1/2 |
| T | $1/2$ | −1 | −1/2 |
| | | Sum | 0 |

So the expected value of the game is zero. Let us turn the game into a two-step game. We flip a fair coin twice. We can represent the game as a tree, as in Figure 9.1. The expectation for this two-step game is zero, and it is a fair game.

*A Modification.* Let's stick with our two-step game but change it slightly.

**1.** If the first toss is a head, switch to a coin where

$$Pr[H] = \frac{3}{4}, \qquad Pr[T] = \frac{1}{4}$$

**2.** If the first toss is a tail, switch to a coin where

$$Pr[H] = \frac{1}{4}, \qquad Pr[T] = \frac{3}{4}$$

We claim that, *as a two-step game,* this is still a fair game. Look at the probability tree in Figure 9.2.

*Caution.* We now come to an important distinction and one of the reasons for including this example. Note that, although this version is fair as a two-step game, it

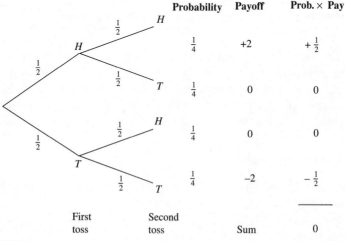

**FIGURE 9.1**
Tree for two-step
toss of a fair coin

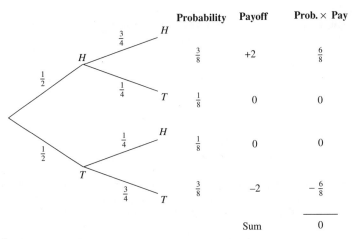

**FIGURE 9.2**
Tree of two-step game involving
switch to unfair coin

clearly is *not* fair as a one-step game. If we are permitted to just watch and not play step 1, and then we play at step 2, we have a clear advantage and a positive expectation. Thus, if we are the player with a payoff of +1 for a head and −1 for a tail, we should play step 2 when the first toss is a head and we should *not* play when the first toss is a tail. To put the matter another way, the second step is not a fair game.

**Comment.** It should be obvious that if we can watch step 1 and play only when that toss is a head, and if we have the "good" coin ($P_R[H] = \frac{3}{4}$), then we have an advantage. The "real world" game of blackjack follows this pattern. Players known as "counters" keep track of the early cards in the deck and then play accordingly. Casinos discourage such practices. (An incident of this sort is featured in the movie *Rain Man*.)

**Summary.** Given a multistep game where the player is permitted to play or sit out any of the steps, the game is fair if and only if all steps are fair.

**Fair is dull.** We do not have to limit ourselves to fair games. Suppose we build a bias into the game. We keep the same fair coin, but we modify the payoff. Thus, we receive $1.05 when the toss is a head and still pay out $1 when the toss is a tail. The tree for the two-step game is shown in Figure 9.3. The expected or average gain for the heads player in the two-step game is five cents per play (per two-step game).

## 9.1.2 The Ho-Lee Model

The first no-arbitrage model for bond prices, presented by T. S. Y. Ho and S. B. Lee in 1986,[1] took the form of a binomial tree.

---

[1]Ho, T. S. Y., and Lee, S. B., "Term Structure Movements and Pricing Interest Rate Contingent Claims," *Journal of Finance*, 41 (December 1986), 1011–1029.

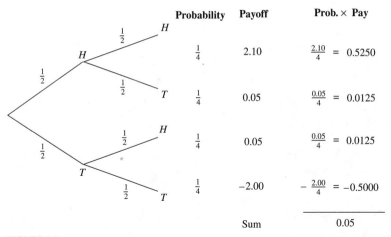

| | Probability | Payoff | Prob. × Pay |
|---|---|---|---|

$H$

$H$    $\frac{1}{2}$

$\frac{1}{2}$

$\frac{1}{2}$

$T$    $\frac{1}{4}$    2.10    $\frac{2.10}{4} = 0.5250$

$\frac{1}{4}$    0.05    $\frac{0.05}{4} = 0.0125$

$H$

$\frac{1}{2}$

$\frac{1}{2}$

$T$

$\frac{1}{2}$

$T$    $\frac{1}{4}$    0.05    $\frac{0.05}{4} = 0.0125$

$\frac{1}{4}$    −2.00    $-\frac{2.00}{4} = -0.5000$

Sum      0.05

**FIGURE 9.3**
Tree for unfair two-step game

The following data should be extracted from the bond market that is being modeled.

1. The volatility for the short rate process is assumed to be a constant, determined by data.
2. The initial yield curve for zero-coupon bonds (ZCBs) gives us the initial forward rates $f(0, t)$.
3. We assume $f(0, t) = f_j$ for $t_j \le t \le t_j + \Delta = t_{j+1}$.

Thus, the $t_j$'s are equally spaced. The Ho-Lee process constructs a tree that matches the volatility, $\sigma$, as well as the initial forward rates, $f(0, t)$, for the intervals $[t_j, t_{j+1}]$.

### Building the Tree Model
In the construction of the tree, we must match both short rate averages (*drifts*) as well as the given short rate volatility. Will this be difficult? Will the order in which we perform these tasks make a difference?

- Since the volatility is constant, it is straightforward to achieve this match.
- To fix the drift, we must adjust an average over many nodes. This task is also easy, and it is best to make it the last step.

Let's look a little closer at the drift question. Suppose you have several numbers,

$$a_1, a_2, \ldots, a_n,$$

whose average is $\bar{a}$. Suppose you wish to change every number so that the new numbers will have the average $A$. This is easy. Just replace

$$a_k \quad \text{with} \quad a_k + (A - \bar{a})$$

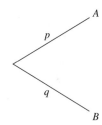

**FIGURE 9.4**
Generic one-step binomial tree
for a random variable

In other words, to fix the average, we *translate* (slide) the entire set up or down. This translation will not affect the volatility, which is why we leave it until the last step.

### Two Facts about Binomials

We are given the one-step binomial tree (or random variable) shown in Figure 9.4. $A$ and $B$ are the two possible values of the random variable $X$. Then,

$$\mu = \text{mean of } X \qquad = pA + qB$$
$$\text{Var}(X) = E((X - \mu)^2) \qquad = (A - B)^2 pq$$

### The Binomial Model

To simplify matters, we will follow the model in Figure 9.5, computing interest rates and growth of principal as $B(\Delta t) = (1 + r\Delta t)B_0$ rather than as $B(\Delta t) = e^{r\Delta t}B_0$. For small values of $\Delta t$, the two methods are almost indistinguishable.

We will construct a model for $B_0 = \$1$. This model will help us evaluate ZCB prices.

1. For the time period $[t_0, t_1]$, the interest rate is $r_0$, the short rate. Thus $B(t_1) = (1 + r_0\Delta t)B_0 = B_1$ since $t_1 - t_0 = \Delta t$.

2. For the period $[t_1, t_2]$ we assume that the short rate takes one of two values, $r_{1,u}$ or $r_{1,d}$, and we use $p = q = 0.5$. From the yield curve, we determine that the interest rate for this period (as predicted by $P_0(t_1, t_2)$) is $f_1$. We wish to match this interest rate, on average, in addition to matching the volatility for the rate. Thus we must satisfy two equations:

$$\frac{r_{1u} + r_{1d}}{2} = f_1 \qquad (a)$$

(so that $r_{1u} + r_{1d} = 2f_1$), and

$$(r_{1u} - r_{1d})^2 = 4\,\Delta t\sigma^2 \qquad (b)$$

This gives us the relation

$$r_{1u} - r_{1d} = 2\sigma\sqrt{\Delta t}$$

**FIGURE 9.5**
Growth of principal under simple interest

$(1 + r_{1u}\Delta t)B_1$
$(1 + r_0\Delta t)\,B_0$
$B_0$
$= B_1$
$t_0$
$t_1$
$(1 + r_{1d}\Delta t)B_1$
$t_2$


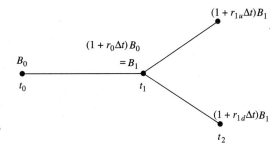

**FIGURE 9.6**
Possible money market values after two binomial steps

Note that $t_{j+1} - t_j = \Delta t$ . We can easily solve (*a*) and (*b*), and we find

$$r_{1u} = f_1 + \sigma\sqrt{\Delta t}$$
$$r_{1d} = f_1 - \sigma\sqrt{\Delta t}$$

Please note the symmetry about the value $f_1$. The money market values are shown in Figure 9.6.

We wish to continue with the construction of the short rate tree. Let us start over again and focus for the moment on "getting the volatility right." In this case the tree becomes very simple, as shown in Figure 9.7.

The tree is just an arithmetic random walk. The volatility along the paths is exactly what we want (and require). However, the average over the nodes (the average interest rate for the period $[t_k, t_{k+1}]$) is $r_0$. We will say that this short rate model, with its simple average, is a "naïve" model.

This is not what we want; the average should be $f_k$, the initial forward rate value. Fortunately, it is easy to rectify this situation. Look at the period $[t_k, t_{k+1}]$. Let $r'_{k,j}$ be the interest rate on *one* branch of the tree for this period.

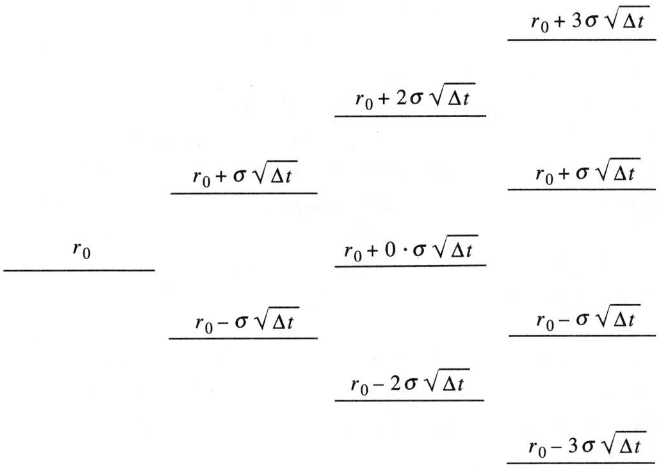

$r_0 + 3\sigma\sqrt{\Delta t}$
$r_0 + 2\sigma\sqrt{\Delta t}$
$r_0 + \sigma\sqrt{\Delta t}$
$r_0 + \sigma\sqrt{\Delta t}$
$r_0$
$r_0 + 0\cdot\sigma\sqrt{\Delta t}$
$r_0 - \sigma\sqrt{\Delta t}$
$r_0 - \sigma\sqrt{\Delta t}$
$r_0 - 2\sigma\sqrt{\Delta t}$
$r_0 - 3\sigma\sqrt{\Delta t}$


**FIGURE 9.7**
"Naïve" short rate tree

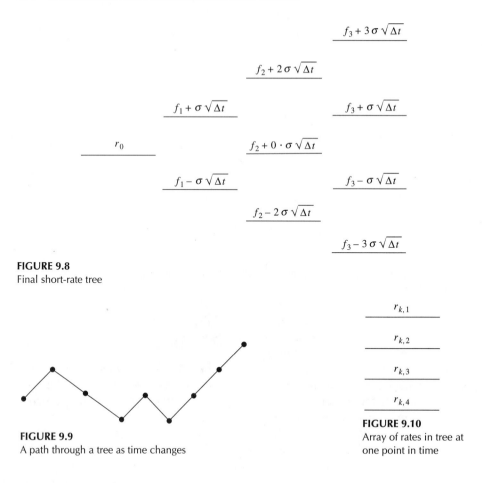

**FIGURE 9.8**
Final short-rate tree

**FIGURE 9.9**
A path through a tree as time changes

**FIGURE 9.10**
Array of rates in tree at
one point in time

Set $r_{k,j} = r'_{k,j} - r_0 + f_k$. With this choice, the average of the $r_{k,j}$'s over $j$ with $k$ fixed (in words, the average of the interest rates over all branches for the period $[t_k, t_{k+1}]$) is $f_k$, the forward rate for that period. So the interest rates on all branches of the new (and final) tree are the $r_{kj}$'s.

Although this adjustment appears complicated when expressed with a formula, it amounts to the simple adjusted tree values given in Figure 9.8.

This construction is a good exercise that requires you to think of the tree both horizontally (as paths, as time changes; see Figure 9.9) and vertically (as a series of rates, with time frozen; see Figure 9.10).

Once the $r_{k,j}$'s have been determined, it is easy to complete the money market price tree, as in Figure 9.11.

**Caution.** Note that although $r_{2,2} = r_{2,3}$, the tree is *not* recombining, and $B_{3,2} \neq B_{3,3}$. Since the tree does not recombine, the number of nodes grows as $2^{k-1}$ (there are $2^{k-1}$ nodes at time $t_k$). This exponential growth creates substantial problems, even for today's computers.

Although this tree construction may appear imposing at first glance, once you have run through an example, such as the following one, it should look simple.

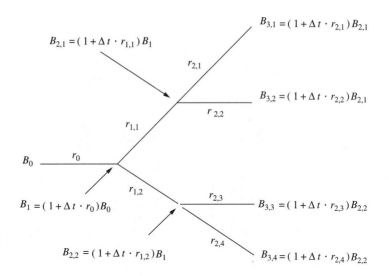

**FIGURE 9.11**
Money market
price tree

---

**Example 1 (Ho-Lee)** We are given the following data:

1. From the yield curve, we determine the forward rates for three-month periods and using recent data, we estimate $\sigma$.

$$r_0 = 0.0485$$
$$f_1 = 0.051$$
$$f_2 = 0.0525$$
$$f_3 = 0.055$$
$$f_4 = 0.055$$

2. $\sigma = 0.01$

Find the money market tree for the first four time periods.

**Step 1.** We begin by finding a "naïve" short rate tree. Thus, $r'_{1u} = r_0 + \sigma\sqrt{\Delta t} = 0.0485 + \sqrt{0.25}(0.01) = 0.0535$. Since $\sigma\sqrt{\Delta t} = \sqrt{0.25}(0.01) = 0.005$, we easily find these short rates, as shown in Figure 9.12. We have used a slightly different format for the tree to emphasize the constancy of the rate over the period $[t_j, t_{j+1}]$.

**Step 2.** We next find the short rate values in the model by adjusting those averages to agree with the initial forward rates, as we described in "Building the Tree Model."

| Time Period | Initial Forward Rate | Adjustment Short Rate – Naïve Average |
|---|---|---|
| $[t_1, t_2]$ | 0.051 | $0.0510 - 0.0485 = 0.0025$ |
| $[t_2, t_3]$ | 0.0525 | $0.0525 - 0.0485 = 0.0040$ |
| $[t_3, t_4]$ | 0.055 | $0.055 - 0.0485 = 0.0065$ |

We add the adjustment to the "naïve" rates and obtain Figure 9.13.

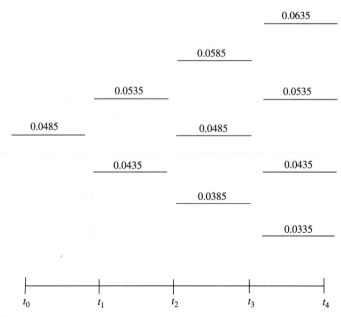

**FIGURE 9.12**
Naïve rate tree

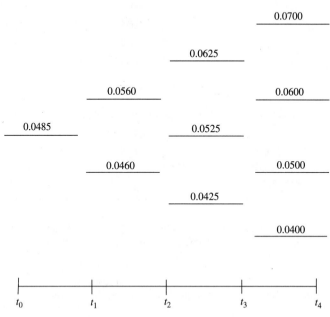

**FIGURE 9.13**
Adjusted short
rate tree

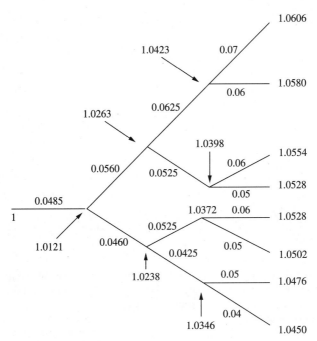

**FIGURE 9.14**
Money market value tree

**Money Market Tree** To complete the example, we compute the money market tree using the formula

$$B_{k+1,j} = (1 + r_{k,j}\Delta t)B_{k,j}$$

with $\Delta t = 0.25$. The result is shown in Figure 9.14.

**Remark.** Note that the money market tree does not recombine.

**Example 2 (Ho-Lee)** We are given the same data as in Example 1. However, we take $\Delta t = 1$ in this example. This time, we wish to compute the ZCB prices for two bonds using Ho and Lee's original method of backward induction; one bond matures at the end of period four and the other bond matures at the end of period three.

We will develop two *bond trees*. The nodes at maturity will have the price of $1, the face value of the bond at maturity. The preceding nodes will contain the discounted bond prices, which are compatible with our short rate tree developed in Example 1.

Each node will have a price determined by the chaining argument of Chapter 3. The unusual feature is that the discounting varies from node to node. Suppose that our short rate tree has the interest rates in Figure 9.15. Then the chaining rule, equation (3.3), applies and

$$P_k = 0.5(1 + r_{k,u}\Delta t)^{-1} \cdot P_{k+1,u} + 0.5(1 + r_{k,d}\Delta t)^{-1} \cdot P_{k+1,d}$$

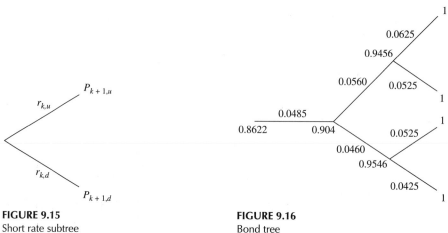

**FIGURE 9.15**
Short rate subtree

**FIGURE 9.16**
Bond tree

So we use this rule to fill in our bond trees. First we take the bond that matures at period 3. The tree is developed as shown in Figure 9.16. The initial bond price is 0.8622, the entry at the root node. In a similar manner, we can compute the tree for the bond maturing at period 4. The details will be left to the reader.

## EXERCISES

1. Given the following forward rates for four-month periods,

$$r_0 = 0.05$$
$$f_1 = 0.0515$$
$$f_2 = 0.052$$
$$f_3 = 0.0525$$
$$\sigma = 0.01$$

   find the money market price tree for the first four periods using the Ho-Lee model.

2. Given the following forward rates for 6-month periods,

$$r_0 = 0.049$$
$$f_1 = 0.050$$
$$f_2 = 0.051$$
$$f_3 = 0.052$$
$$\sigma = 0.011$$

   find the money market price tree for the first four periods using the Ho-Lee model.

## 9.2   A BINOMIAL VASICEK MODEL: A MEAN REVERSION MODEL

In this section we will take the Ho-Lee model from the previous section (Figure 9.17) and add an extra step in the construction. This step introduces **mean reversion** into

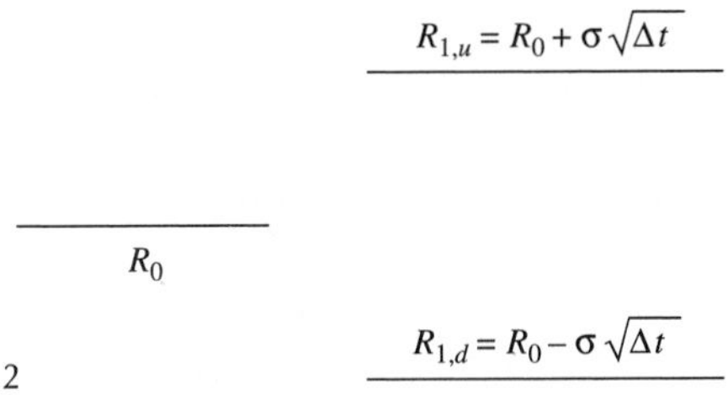

$$R_{1,u} = R_0 + \sigma\sqrt{\Delta t}$$

$$R_0$$

**FIGURE 9.17**
Rate tree from Section 9.1.2

$$R_{1,d} = R_0 - \sigma\sqrt{\Delta t}$$

the model. Thus this extra step "pushes" the interest rates back toward the mean or average rate.

Recall that the expression or model for $r$ in the Vasicek model is

$$dr = a(b - r)dt + \sigma dB$$

We will use the first term on the right to implement mean reversion.

We assume that we know

1. All forward rates $f_j$ for the periods $[t_j, t_{j+1}]j = 0, \ldots, N - 1$
2. $a, \sigma$
3. $b$, the average or mean interest rate for the period $[t_0, t_N]$

We will develop this model using the method of mathematical induction in two steps. The first applies to the base case: the first time period. The second applies to the next time period after the preceding time period had been computed.

## 9.2.1   The Base Case

**Step 1. The Reversion Step**
We first adjust $R_0$ before extending the tree to

$$R_{1,u} = R_0 + \sigma\sqrt{\Delta t}$$
$$R_{1,d} = R_0 - \sigma\sqrt{\Delta t}$$

Thus we replace $R_0$ by

$$R_0^r = R_0 + a(b - R_0)\Delta t$$

The superscript letter $r$ stands for reversion. This step implements the mean reversion.

**Step 2. Extending the Tree**
We define

$$R_{1,u}^{\#} = R_0^r + \sigma\sqrt{\Delta t}$$
$$R_{1,d}^{\#} = R_0^r - \sigma\sqrt{\Delta t}$$

as shown in Figure 9.18. This step guarantees the correct volatility.

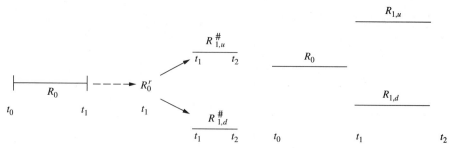

**FIGURE 9.18**
Extending the tree from $t_1$ to $t_2$ in the Vasicek model

**FIGURE 9.19**
Interest rate tree carried to time $t_2$

## Step 3. Matching the Forward Rate

The forward rate does not match the average of $R_{1,u}$ and $R_{1,d}$ at this point, but that is easy to fix. Set

$$R_{1,u} = R_{1,u}^{\#} - \left(\frac{R_{1,u}^{\#} + R_{1,d}^{\#}}{2}\right) + R_1$$

and

$$R_{1,d} = R_{1,d}^{\#} - \left(\frac{R_{1,u}^{\#} + R_{1,d}^{\#}}{2}\right) + R_1$$

This completes the calculation for the second stage of the interest rate tree, as shown in Figure 9.19.

The next stage of this construction ensures that the paths will have the correct volatilities and that the average of the tree interest rates for a fixed period $[t_k, t_{k+1}]$ will match the forward rate $f_k$.

## 9.2.2   The General Induction Step

We assume we have found the forward interest rate $R(t_n, j)$. This is the interest rate on path $j$ at time $t_n$. At this time, there are $2^{n-1}$ nodes.

### Step 1. Reversion Step

We replace $R(t_n, j)$ by $R^r(t_n, j) = R(t_n, j) + a(b - R(t_n, j))\Delta t$.

### Step 2. Extending the Tree

We set

$$R^{\#}(t_{n+1}, j^+) = R^r(t_n, j) + \sigma \Delta t$$

and

$$R^{\#}(t_{n+1}, j^-) = R^r(t_n, j) - \sigma \Delta t$$

for $j = 1$ to $2^{n-1}$. Thus, there are now $2^n$ nodes at time $t_{n+1}$. The term $\sigma \Delta t$ ensures that the volatility along the path is $rj$ matching the real world value.

## Step 3. Matching the Forward Rate

$R_{t_{n+1}}$, which we are given, is the short rate for the model for the period $[t_{n+1}, t_{n+2}]$. We want the $R(t_{n+1}, j)$'s, which we construct, to reflect this rate. We set $\bar{R}_{t_{n+1}} = 2^{-n} \sum_{j=1}^{2n} R^{\#}(t_{n+1}, j)$. To complete the process, we set $R(t_{n+1}, j) = R^{\#}(t_{n+1}, j) - \bar{R}_{t_{n+1}} + f_{t_{n+1}}$. This last definition may look complicated, but it is really simple. The average of the first two terms on the right-hand side is 0. Thus, if we average $R(t_{n+1}, j)$ over the nodes (over $j$), we just get $f_{t_{n+1}}$, the forward rate for the period $[t_{n+1}, t_{n+2}]$.

## Summary

All the forward rates and volatilities for the model match up, because they were constructed to do so.

---

### Example: Reversion Model

1. From the yield curve from Examples 1 and 2 in Section 9.1, we determine forward rates for three-month periods to be

$$r_0 = 0.0485$$
$$f_1 = 0.051$$
$$f_2 = 0.0525$$
$$f_3 = 0.055$$
$$R_4 = 0.057$$

2. We are also given

$$a = 0.4$$
$$\Delta t = \tfrac{1}{4}$$
$$b = 0.056$$
$$\sigma = 0.01$$

Find the money market tree using the reversion model. We start with $R_0 = 0.0485$. Hence $R_0^r = 0.04925$. The tree extends to $t_2$ in Figure 9.20, to $t_3$ in Figure 9.21, and to $t_4$ in Figure 9.22. The rate tree is summarized in Figure 9.23, and from it the money market tree in Figure 9.24 is computed.

---

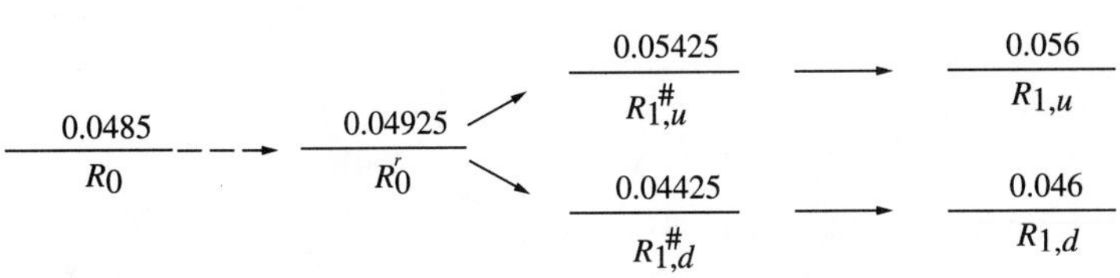

**FIGURE 9.20**
Rate tree, $t_0$ to $t_2$

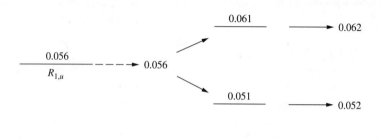

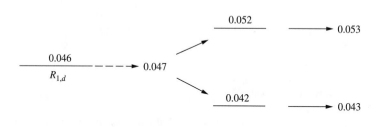

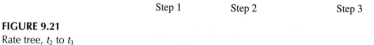

Reversion

Step 1          Step 2          Step 3

**FIGURE 9.21**
Rate tree, $t_2$ to $t_3$

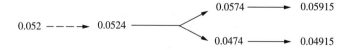

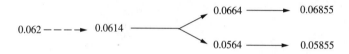

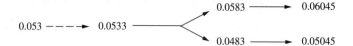

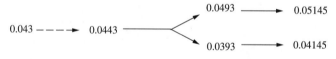

Step 1          Step 2          Step 3

**FIGURE 9.22**
Rate tree, $t_3$ to $t_4$

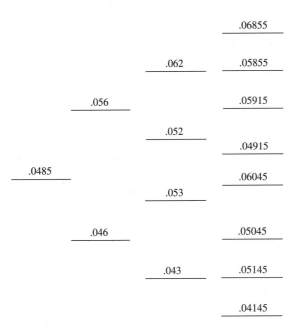

**FIGURE 9.23**
Summary of rate tree

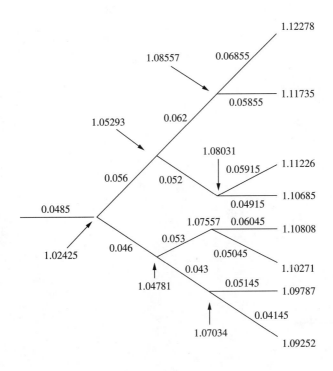

**FIGURE 9.24**

## EXERCISES

1. Use the following rate data to compute the money market tree for the mean reversion model.

- The forward rates for three-month periods are

$$f_0 = 0.055$$
$$f_1 = 0.058$$
$$f_2 = 0.0625$$
$$f_3 = 0.055$$
$$f_4 = 0.050$$

- We are also given

$$a = 0.5$$
$$\Delta t = 1/4$$
$$b = 0.03$$
$$\sigma = 0.02$$

# CHAPTER
# 10

## CURRENCY MARKETS AND FOREIGN EXCHANGE RISK

*We're not in it for the money.*
*We're in it for the fun,*
*Money's just how we keep score.*

George Goodman, *The Wheeler Dealers*

The foreign exchange (FX), or currency, markets are the largest of all financial markets. Global trading is roughly two trillion dollars per day. The leading FX trading centers are London, New York, and Tokyo, in that order.

## 10.1 THE MECHANICS OF TRADING

A foreign exchange transaction really *is* an exchange. One delivers marks and receives pounds, for example. There are two pricing conventions:

- *American convention:* Number of units of domestic currency = 1 unit of foreign currency. Thus, for example, in the American convention, if the dollar is the domestic currency and the mark is the foreign currency, the quote would be

$$0.5434 \text{ dollars } = 1 \text{ mark}$$

- *European convention:* Number of units of foreign currency = 1 unit of domestic currency. Assuming that the dollar is still the domestic currency as in the previous example then, in the European convention, the quote would be

$$1.8404 \text{ marks } = 1 \text{ dollar}$$

Note that the exchange rates in the two conventions are reciprocals, so, given one, it is easy to determine the other.

## EXERCISES

1. The *Wall Street Journal* for May 16, 2000, lists the following (spot) exchange rates.

| American quote: | | |
|---|---|---|
| One dollar | = 2.1451 | German marks |
| One dollar | = 7.1945 | French francs |
| One dollar | = 8.1829 | Danish kroner |
| One dollar | = 0.6638 | British pounds |
| One dollar | = 182.49 | Spanish pesetas |
| One dollar | = 109.48 | Japanese yen |

| European quote: | | |
|---|---|---|
| One Belgium franc | = 0.0226 | dollars |
| One Swiss franc | = 0.5859 | dollars |
| One Swedish krona | = 0.1103 | dollars |
| One Dutch guilder | = 0.4137 | dollars |
| One Brazilian real | = 0.5484 | dollars |
| One Canadian dollar | = 0.6424 | dollars |

We are considering the U.S. dollar to be the domestic currency and all other currencies to be the foreign currency. (a) Convert all American quotes to European quotes. (b) Convert all European quotes to American quotes.

2. Using the information in Exercise 1 it is possible to determine currency equivalents for other currencies. For example,

$$2.1451 \text{ German marks } = \text{ one dollar } = 7.1945 \text{ French francs}$$

Thus one mark $= 7.1945/2.1451 = 3.3539$ francs.
   Find currency equivalents in the following cases:

   **(a)**  marks / Danish kroner
   **(b)**  Swiss franc / guilder
   **(c)**  pound / peseta
   **(d)**  yen / mark
   **(e)**  Danish kroner / pound
   **(f)**  French franc / yen
   **(g)**  Canadian dollar / British pound
   **(h)**  real / guilder
   **(i)**  Swedish krona / Swiss franc
   **(j)**  Belgium franc / French franc

## 10.2  CURRENCY FORWARDS: INTEREST RATE PARITY

As described in Chapter 1, a forward contract is an agreement between two parties to buy/sell a commodity (oil, grains, currency, stock indices) at an agreed-on time (in the future) for an agreed-on sum. No money changes hands at this point. For example, party A may agree to buy one million barrels of oil on June 17, 2000, at the price of \$21 per barrel from party B.

In a typical forward contract, a company would exchange marks for dollars. The settlement would take place at some date in the future. Usually the counterparty is a bank. Note that in such a trade the company is *buying* marks and *selling* dollars. How is the price computed? In effect, the price is determined by the interest rates in the two countries plus the (nominal) exchange rate.

Suppose an American bank is doing the trade with ABC Company. The bank will deliver one mark to ABC Company in the future. In return, ABC Company delivers a certain number of dollars at that time. Let

$$S_0 \text{ dollars} = 1 \text{ mark ( at } t = 0)$$

$$R_D = \text{riskless U.S. interest rate}$$

$$R_F = \text{riskless German interest rate}$$

At time $t = 0$ the bank borrows $S_0 e^{-R_F \tau}$ dollars. These are converted to $e^{-R_F \tau}$ marks:

|  | Dollars |  | Marks |
|---|---|---|---|
| $t = 0$ | $S_0 e^{-R_F \tau}$ | $\longrightarrow$ | $e^{-R_F \tau}$ |
|  | $\downarrow$ |  | $\downarrow$ |
|  |  |  | Marks |
| $t = \tau$ | $S_0 e^{-R_F \tau} \cdot e^{R_D \tau}$ | $\longleftrightarrow$ | 1 |

The $e^{-R_F \tau}$ marks are invested at the foreign (German) rate. At $t = \tau$ they have grown to one mark, which is delivered to ABC. To break even, the bank must receive

$$S_0 e^{(R_D - R_F)\tau} \text{ dollars} \tag{10.1}$$

at time $\tau$.

Equation (10.1) is Keynes' famous **interest parity formula**. In general, $M$ units of foreign currency are exchanged for

$$M S_0 e^{(R_D - R_F)\tau} \tag{10.2}$$

units of domestic currency, where

$$R_D = \text{domestic interest rate}$$

$$R_F = \text{foreign interest rate}$$

$$\tau = \text{settlement date}$$

$$S_0 \text{ units of domestic currency} = 1 \text{ unit of foreign currency}$$

Note that if any price other than (10.2) is offered (on either the buy or sell side), then a riskless arbitrage opportunity is available.

**Example**  We are given the following data:

$$1.5065 \text{ U.S. dollars} = 1 \text{ British pound}$$

$$\text{Short-term U.S. interest rate} = 0.062$$

$$\text{Short-term British interest rate} = 0.06$$

Find the price in dollars for a forward contract that delivers 1 million pounds 3 months from today.

**Solution:**  We use the interest rate parity formula,

$$F = S_0 M e^{(R_D - R_F)\tau}$$

Thus,

$$F = 1.5065 \times 1,000,000 e^{(0.062 - 0.06) \times 0.25}$$
$$= \$1,507,253.40$$

## EXERCISES

**1.** We are given the following data:

$$\text{Short-term U.S. interest rate} = 0.062$$

$$\text{Short-term German interest rate} = 0.037$$

$$0.462 \text{ dollars} = 1 \text{ mark}$$

Find the forward price in dollars for delivery of 1 mark six months from today.

**2.** Given the following data,

$$\text{Short-term U.S. interest rate} = 0.062$$

$$\text{Short-term French interest rate} = 0.039$$

$$0.1390 \text{ dollars} = 1 \text{ franc}$$

Find the forward price in dollars for delivery of 1 franc three months from today.

**3.** Given the following data,

$$\text{Short-term U.S. interest rate} = 0.062$$

$$\text{Short-term Japanese interest rate} = 0.02$$

$$0.009134 \text{ dollars} = 1 \text{ yen}$$

Find the forward price in dollars for 1 yen to be delivered two months from today.

**4.** Given the following data,

$$0.5484 \text{ dollars} = \text{ one Brazilian real}$$

$$\text{U.S. short-term interest rate} = 5.95\%$$

$$\text{Brazilian short-term interest rate} = 6.1\%$$

Find the price in dollars for a forward contract that delivers 10 million reals one year from today.

## 10.3   FOREIGN CURRENCY OPTIONS

### 10.3.1   The Garman-Kohlhagen Formula[1]

We wish to price an option on a foreign currency. We adopt the following notation:

$S_0$ = spot price of foreign currency (domestic units for one foreign unit)

$F$ = forward price

$K$ = exercise price (domestic units for one foreign unit)

$T$ = time to maturity

$C(S, T)$ = price of FX call option (domestic units for one foreign unit)

$R_D$ = domestic riskless rate

$R_F$ = foreign riskless rate

$\sigma$ = volatility of the spot currency price

$\mu$ = drift of the spot currency price

$\alpha$ = expected rate of return on a security

$\delta$ = standard deviation of the security rate of return

We make the following assumptions:

1. The spot currency price is modeled by $dS = \mu S dt + \sigma S dB$, where $B$ is a standard Brownian motion.
2. Option prices are functions of $S, T$ only.
3. Interest rates ($R_D$ and $R_F$) are constant.
4. Constancy of **market price of risk**:

$$\frac{\alpha_i - R_D}{\delta_i} = \lambda \text{ for all securities } i \tag{10.3}$$

Thus,

$$\frac{\mu + R_F - R_D}{\sigma} = \lambda$$

and

$$\frac{\alpha_C - R_D}{\delta_C} = \lambda \text{ where } C = C(S, T)$$

---

[1]Garman, M., and Kohlhagen, S., "Foreign Currency Option Values," *Journal of International Money and Finance*, 2(1983), pp. 231–238.

Expand $C$ in a Taylor series. Thus,

$$dC = -\frac{\partial C}{\partial T}dt + \frac{\partial C}{\partial S}dS + \frac{1}{2}\frac{\partial^2 C}{\partial S^2}dS^2 + \cdots \qquad (10.4)$$

$$dS = \mu S dt + \sigma S dB \qquad (10.5)$$

Using Ito's lemma we see that

$$dS^2 \cong \sigma^2 S^2 dt \qquad (10.6)$$

Substitute (10.5) and (10.6) in (10.4) to obtain

$$dC = -\frac{\partial C}{\partial T}dt + \left[\mu S \frac{\partial C}{\partial S} + \frac{1}{2}\sigma^2 S^2 \frac{\partial^2 C}{\partial S^2}\right]dt + \sigma S \frac{\partial C}{\partial S}dB \qquad (10.7)$$

But

$$dC = \alpha_C C dt + \delta_C C dB \qquad (10.8)$$

Equating (10.7) and (10.8), we see that

$$\alpha_C C = -\frac{\partial C}{\partial T} + \mu S \frac{\partial C}{\partial S} + \frac{1}{2}\sigma^2 S^2 \frac{\partial^2 C}{\alpha S} \qquad (10.9)$$

and

$$\delta_C C = \sigma S \frac{\partial C}{\partial S} \qquad (10.10)$$

We can rewrite (10.3) as

$$\frac{\alpha_C C - R_D C}{\delta_C C} = \lambda \qquad (10.11)$$

Substituting (10.9) and (10.10) in (10.11), we see that

$$\frac{-\dfrac{\partial C}{\partial T} + \mu S \dfrac{\partial C}{\partial S} + \dfrac{1}{2}\sigma^2 S^2 \dfrac{\partial^2 C}{\partial S^2} - R_D C}{\sigma S \dfrac{\partial C}{\partial S}} = \lambda = \frac{(\mu + R_F) - R_D}{\sigma}$$

Cancel the $\sigma$'s in the denominator. Then cross-multiply and finally rearrange to end up with

$$-\frac{\partial C}{\partial T} + (R_D S - R_F S)\frac{\partial C}{\partial S} + \frac{1}{2}\sigma^2 S^2 \frac{\partial^2 C}{\partial S^2} = R_D C$$

The differential equation should look very familiar. As it stands, it is the Black-Scholes equation for an option on a stock paying a dividend, and if we set $R_F = 0$, it is the Black-Scholes equation for a non–dividend-paying stock.

We note the usual boundary conditions:

1. $C(S, 0) = [S_0 - K]^+$
2. $C(0, T) = 0$
3. $\lim_{S \to \infty} \frac{C(S,T)}{S} = 1$

The solution to the differential equation is

$$C(S, T) = e^{-R_F T} SN(d_1) - e^{-R_D T} KN(d_2)$$

where $N$ is the normal distribution and

$$d_1 = \frac{\ln(S/K) + (R_D - R_F + \sigma^2/2)T}{\sigma\sqrt{T}}$$

and

$$d_2 = \frac{\ln(S/K) + (R_D - R_F - \sigma^2/2)T}{\sigma\sqrt{T}}$$

The expression for $C$ is quite a bit cleaner if we use the forward price. Thus, invoking the interest parity formula, $F = Se^{(R_D - R_F)T}$, we find

$$C(S, T) = [FN(d_1) - KN(d_2)]e^{-R_D T}$$

where

$$d_1 = \frac{\ln(F/K) + \sigma^2 T/2}{\sigma\sqrt{T}}$$

and

$$d_2 = \frac{\ln(F/K) - (\sigma^2 T/2)}{\sigma\sqrt{T}}$$

## 10.3.2   Put-Call Parity for Currency Options

Just as there is a put-call parity formula for stock options, there is an analogous formula for currency options. We start with a fixed pair of domestic and foreign currencies.

We specify a call $C$ to deliver one unit of foreign currency at time $T$ in the future. In return, the holder of the call pays $K$ units of domestic currency to the counterparty at delivery time. Thus $K$ is the strike price (in domestic units for one foreign unit).

Let $P$ be a put to sell one unit of foreign currency at the same strike ($K$ units of domestic currency), and with the same expiration date $T$. We assume the put and the call are priced in domestic currency. Let $S_0$ be the spot price today ($S_0$ units of domestic = one unit foreign).

Our program is as follows:

**1.** We buy one put.
**2.** We sell one call.
**3.** We borrow $S_0 e^{-R_F T}$ units of domestic.
**4.** We exchange it for $e^{-R_F T}$ units of foreign.
**5.** We invest this sum at rate $R_F$.

Our total outlay of cash is

$$P + S_0 e^{-R_F T} - C$$

At time $T$ we must settle up. Our $e^{-R_F T}$ of foreign has appreciated to one unit of foreign. At time $T$, the spot price is $S_T$.

- If $S_T \geq K$, then the call will be exercised against us, and we deliver the one unit of foreign in our possession and receive $K$ domestic.
- If $S_T < K$, then we deliver the one unit of foreign to the seller of the put and receive $K$ units of domestic.

So, no matter what spot price obtains at $T$, we deliver one unit of foreign and receive $K$ units of domestic. Because we invested

$$P - C + S_0 e^{-R_F T} \text{ (domestic)}$$

at time $t = 0$, we expect that

$$(P - C + S_0 e^{-R_F T})e^{R_D T} = K$$

whence

$$C - P = S_0 e^{-R_F T} - e^{-R_D T} K$$

Note that this transaction is completely deterministic, so if any price differential other than $C - P$ is available, then it is possible to make a riskless profit through arbitrage.

## 10.4   GUARANTEED EXCHANGE RATES AND QUANTOS

An investor who purchases a foreign stock or a foreign stock index is exposed to two risks. First, the stock or the index may perform adversely. Second, even when the stock or index performs admirably, the investor is at risk from the currency exchange. There are several ways the investor can hedge such risk.

Suppose an investor agrees to buy stock $S$ at time $T$ for a forward price $K$ where $S$ and $K$ are denominated in foreign currency. Let "one unit of foreign equal $X_T$ units of domestic" be the spot rate at time $T$.

*Unwinding the Trade.* On the settlement date, under standard conditions, the investor would receive

$$(S_T - K)X_T \qquad \text{in domestic} \tag{10.12}$$

However, it is possible to enter into a **Guaranteed Exchange Rate (GER) contract** when the parties agree on a preassigned exchange rate $X^{\#}$ independent of $X_T$. Under the circumstances just described, when the trade is unwound at time $T$, the investor receives

$$(S_T - K)X^{\#} \qquad \text{in domestic} \tag{10.13}$$

How should this forward contract be priced? The dealer wishes to choose $K$ sufficiently large so as to protect against loss and make a profit to compensate for the risk.

*Historical Note.* Nikkei Stock Index futures with preset exchange rates have traded on the Singapore International Monetary Exchange (SIMEX) since 1986. Nikkei Index futures with preset exchange rates are also traded on the American Stock Exchange. Such instruments are also known as **quantos**.

Before trying to price a GER on a stock, let us look at the hedging of a bond against foreign exchange risk.

## 10.4.1   The Bond Hedge

Suppose a bond is redeemed at maturity (time $t = T$) for $Q$ units of foreign currency. To hedge the position, you enter into a forward contract maturing at time $T$. The contract specifies that you pay the counterparty $Q$ units of foreign at $T$ and receive in return

$$D = S_0 e^{(R_D - R_F)T} Q \text{ units of domestic} \tag{10.14}$$

as determined by the interest parity formula, where $S_0$ units of domestic $= 1$ unit of foreign (the spot price at $t = 0$), $R_D$ = domestic riskless rate, $R_F$ = foreign riskless rate.

Although this appears straightforward, the hedging process is considerably more complicated in practice. Instead of entering into a forward contract for the period $[0, T]$, one might wish to use a series of shorter periods.

Let us spell out the differences in detail.

**Definition.** A foreign currency is said to be a **premium** currency if its interest rate is *lower* than that of the domestic currency. A foreign currency is said to be a **discount** currency if its interest rate is *higher* than that of the domestic currency.

Which is preferable, one long-term forward hedge or a series of rolling short-term hedges? We present a summary in Table 10.1. The spread refers to the quantity $|R_D - R_F|$.

TABLE 10.1
**Guidelines for bond hedging with exchange rates**

|  | Premium Currency | Discount Currency |
|---|---|---|
| Spread narrows | Long-term better | Short-term better |
| Spread widens | Short-term better | Long-term better |

To check this out, take the entry in the upper left corner. If the spread narrows, the factor $R_D - R_F$ is less than 0 but is smaller in absolute value, so you will have to pay more on each rollover.

### 10.4.2   Pricing the GER Forward on a Stock

Let us turn to the case of the stock valued in a foreign currency, with which we began this section. There is a huge difference between a stock and a bond. In the case of a bond, we know the final payment at maturity and can plan accordingly. In the case of a stock (or stock index), we do not know its value even one year in advance. This means we cannot simply buy a forward.

We wish to value a GER contract or forward on a foreign stock. Assume the stock is priced in francs and the domestic currency is dollars. Let

$$X_t = \text{dollar value of 1 franc at time } t$$

We take as our model for $X$

$$dX = r_X X dt + \sigma_X X dB \qquad (10.15)$$

Our model for the stock price is

$$dS = r_S S dt + \sigma_S S dB \text{ (in francs)} \qquad (10.16)$$

At this point $r_S$ is unknown, but we can find it in terms of other drift rates. The factor $r_S$ is the crucial link for evaluating a GER forward on $S(T)$.

Let us list the notation we will use:

$$S(t) = \text{value of stock in francs at time } t$$
$$r_D = \text{U.S. riskless interest rate}$$
$$r_F = \text{French riskless interest rate}$$
$$T = \text{time for delivery of the stock}$$
$$K = \text{stock's delivery price in francs}$$
$$\sigma_{XS} = \text{covariance for the stock price (in francs)}$$
$$\text{and francs' price (in dollars).}$$

We assume that the stock pays no dividend.

Before going further, let us step back for a moment.

## The Big Picture and Strategic Intent

We will determine the expected value $K$ of the stock in francs at maturity ($t = T$). The GER contract then requires the seller to pay or receive the difference

$$S_T - K$$

at time $T$ in dollars. The quantity ($S_T - K$) is denominated in francs. What exchange rate $X_R$ is used? The exchange rate $X_R$ is determined by the counterparts at time $t = 0$. But $X_R$ need not have any connection with existing or future spot rates. In this respect the GER contract strikes one as a bit strange. It resembles a bet where $X_R$ plays the role of the bet size.

## A Digression

Think back to our discussion of Black-Scholes option pricing. Recall that our stock model was

$$dS = \mu S dt + \sigma S dB$$

but the solution to this stochastic differential equation had the form

$$S(t) = S_0 \exp[(\mu - \sigma^2/2)t + \sigma B_t]$$

Where did the strange term $\sigma^2/2$ come from? The answer is that the term $e^{\sigma B_t}$ is not driftless. The $e^{\sigma^2 t/2}$ appears in the equation for $S(t)$ to balance out the drift. The expression

$$e^{-\sigma^2 t/2} \cdot e^{\sigma B_t}$$

is driftless. All the drift is now embedded in the term $e^{\mu t}$.

Since $t$ is fixed, in the following discussion we will write $B_t$ as $\sqrt{t}Z$ and will write $S(t)$ and $X(t)$ in the form

$$S(t) = S_0 e^{r_S t} \cdot e^{-\sigma_S^2 t/2} \cdot e^{\sqrt{t}\sigma_S Z} \qquad (10.17)$$

and

$$X(t) = X_0 e^{r_X t} \cdot e^{-\sigma_X^2 t/2} \cdot e^{\sqrt{t}\sigma_X Z'}$$

so we may easily manipulate drifts. The random terms, $Z$ and $Z'$, are two (possibly correlated) standard normals. We have followed the presentation of Derman, Karasiinski, and Becker[2] for parts of our derivation.

## Growth Rate of the Stock in Dollars

Two factors contribute to the growth rate or drift of the stock in dollars: its growth in francs and the exchange rate. We are using a model where the (total) drift of *any* asset is equal to $r_D$. Once we compute the stock drift, we will set it equal to $r_D$.

---

[2]Derman, E., Karasiinski, P., and Becker, J., "Understanding Guaranteed Exchange Rate Contracts in Foreign Stock Investments," in De Rosa, David, ed., *Currency Derivatives*, Wiley, New York, 1998, pp. 329–339.

First, note that a direct investment in francs is another asset. After the investment gains interest in a French bank and the proceeds are converted to dollars, we see a drift of $r_F + r_X$. Hence,

$$r_D = r_F + r_X \tag{10.18}$$

Now let us find $E[S(t)X(T)]$ to identify the stock drift in dollars. Note that $S(t)$ and $X(t)$ are not independent. Using equation (10.17), we have

$$E[SX] = S_0 X_0 e^{(r_S + r_X)t} \cdot e^{-(\sigma_S^2 + \sigma_X^2)t/2} \cdot E[e^{\sqrt{t}(\sigma_S Z + \sigma_X Z')}]$$

But the random quantity

$$\sigma_S Z + \sigma_X Z'$$

is normal, and its variance is $\sigma_S^2 + 2\sigma_{SX} + \sigma_X^2$. We can use the same rule for $E[e^{\sigma Z}]$ as before to obtain

$$E[\exp\{\sqrt{t}(\sigma_S Z + \sigma_X Z')\}] = \exp[t(\sigma_S^2 + 2\sigma_{SX} + \sigma_X^2)/2]$$

We then obtain the equation

$$E[S(T)X(T)] = \exp[(r_S + r_X + \sigma_{XS})t]S(0)X(0) \tag{10.19}$$

This shows that the drift for the stock in dollars is

$$r_S + r_X + \sigma_{XS}$$

Since any asset drift is equal to $r_D$, we equate this to the left-hand side of equation (10.18):

$$r_F + r_X = r_S + r_X + \sigma_{XS} \tag{10.20}$$

We simplify (10.20) to arrive at

$$r_F = r_S + \sigma_{XS} \tag{10.21}$$

Thus

$$E[S(T)] = S_0 e^{r_S T} = S_0 e^{(r_F - \sigma_{XS})T} \tag{10.22}$$

We now set

$$K = E[S(T)] = S_0 e^{(r_F - \sigma_{XS})T} \tag{10.23}$$

Note that the expected value of the stock at time $T$, in dollars, at the agreed-on exchange rate is

$$E[X_R S(T)] = X_R S_0 e^{(r_F - \sigma_{XS})T} \tag{10.24}$$

Thus

$$E[X_R S(T) - X_R K] = 0$$

The contract is a *fair game* (see Section 9.1.1) in the sense of expected value. At settlement time $T$, the holder of the contract delivers $KX_R$ dollars to the seller of the contract and receives $S_T X_R$ dollars in return. In practice, the seller pays or receives from the holder of the forward the amount

$$(S(T) - K)X_R \text{ dollars}$$

The purchaser of a GER forward on a stock (or stock index) is insured against any currency rate "surprises."

### 10.4.3   Pricing the GER Put or Call Option

To price a GER put or call on a foreign stock, we can follow the procedure we used earlier for domestic stocks.

Let

$S(t)$ = price of stock in francs at time $t$

$K$ = strike price in francs

$r_D$ = riskless U.S. interest rate

$r_F$ = riskless French interest rate

$T$ = time to maturity

$\sigma_S$ = volatility of stock

$\sigma_{XS}$ = covariance for the stock price (in francs) and franc price (in dollars)

$X_R$ = agreed-on exchange rate (number of dollars equals one franc)

To price a European call on a stock paying no dividends, we must find the expected value of

$$\max[S(T) - K, 0]X_R \tag{10.25}$$

The price of the call is

$$E[\max[S(T) - K, 0]]X_R e^{-r_D T} \tag{10.26}$$

Thus,

$$C = [S(T)e^{(r_F - \sigma_{XS})T} N(d_1) - KN(d_2)]X_R e^{-r_D T} \tag{10.27}$$

where

$$d_1 = \frac{\ln(S(0)/K) + (r_F - \sigma_{XS} + \sigma_S^2/2)T}{\sigma_S \sqrt{T}}$$

and

$$d_2 = \frac{\ln(S(0)/K) + (r_F - \sigma_{XS} - \sigma_S^2/2)T}{\sigma_S \sqrt{T}}$$

## 10.5 TO HEDGE OR NOT TO HEDGE—AND HOW MUCH

There are a variety of opinions on hedging currency risk. There seems to be general agreement that fixed income investments should be hedged.[3,4,5] For equities and stock indexes the situation is less clear. The literature contains several excellent discussions.[6,7,8,9]

---

[3]Black, F., "Universal Hedging: How to Optimize Currency Risk and Reward in International Equity Portfolios," *Financial Analysts' Journal* (July/August 1989), pp. 16–22.

[4]Black, F., "Equilibrium Exchange Rate Hedging," *Journal of Finance*, 45, 3 (July 1990) pp. 899–907.

[5]Perold, A., and Schulman, E., "The Free Lunch on Currency Hedging: Implications for Investment Policy and Performance Standards," *Financial Analysts Journal* (May/June 1988), pp. 46–50.

[6]Gastineau, G., "The Currency Hedging Decision: A Search for Synthesis in Asset Allocation," *Financial Analysts' Journal*, 1997.

[7]De Rosa, David, *Managing Foreign Risk*, Irwin, Chicago, 1996.

[8]Froot, K., "Currency Hedging over Long Horizons," Working Paper No. 4355, *National Board of Economic Research*, May 1993.

[9]Piros, C., "The Perfect Hedge: To Quanto or Not to Quanto,"in De Rosa, David, ed., *Currency Derivatives*, Wiley, New York, 1998, pp. 340–353.

# CHAPTER
# 11

# INTERNATIONAL
# POLITICAL
# RISK ANALYSIS

*The fools are dancing,*
*But the greater fools are watching*

Chinese Proverb

## 11.1 INTRODUCTION

The collapse of the Thai baht in the summer of 1997 signaled the outbreak of a series of financial crises that swept across Southeast Asia. Though the crisis was initially concentrated in that region, emerging markets around the globe soon experienced severe economic problems. From Indonesia to Brazil, financial collapse led to real decreases in the standard of living for millions of people. The effects of the crisis were not limited to emerging markets, however. On Wall Street, many financial institutions suffered significant losses due to emerging market positions. Perhaps most notably, Long Term Capital Management (LTCM), a money manager with two Nobel laureates on its payroll, lost billions on emerging market investments when the Russian ruble collapsed in August 1998. U.S. Federal Reserve officials, concerned that an LTCM bankruptcy would spark a market panic, engineered a $3.6 billion restructuring of the company by a consortium of 14 Wall Street banks.

Some economists argued that the emerging markets crisis would have a strong impact on the global economy. Economic forecasting models showed that the troubles in Asia would shave anywhere from a half percentage point to a full point off the growth rate of the U.S. economy in 1998.[1] Laura D'Andrea Tyson, President Clinton's economic adviser during his first term, wrote:

> The Asian financial crisis is one of the gravest of modern times. It holds significant contractionary risks not just for the Asian economies, but for the entire global economy including the United States. It also threatens the geopolitical stability of a region of paramount importance to American security interests.

The turmoil created by these emerging market crises highlights the risks inherent in international investment. Though globalization offers immense opportunities for economic development, it also carries with it certain risks. Managing exposure to international risks is becoming an increasingly important task for financial institutions. The five sections of this chapter (1) provide an overview of the types of risks that investors face, (2) describe techniques employed to manage those risks, (3) describe country risk derivatives in some detail, (4) introduce methods of analyzing international political risk, and (5) explain two types of bond pricing models that can be used to assess political risk.

## 11.2   TYPES OF INTERNATIONAL RISKS

Risk is the possibility of suffering harm or loss, and international investors must cope with two different types of risk:

1. Commercial or market risk
2. Political risk

Investors and entrepreneurs bear market risk in order to earn a return from their initial investment. There is a positive relationship between market risk and potential return, since higher risk generally implies the possibility of a higher return. For example, if you invest in a small Internet startup company instead of a well-established large corporation, you may have the opportunity to earn a higher return on your investment. However, there is also a higher chance that you will lose money than if you invested in a large established company.

Political risk is the other type of risk borne by the international investor.

### 11.2.1   Political Risk

Political risk is the possibility of adverse impact on commercial activities due to actions taken by a government or a political organization. A key difference between market risk and political risk is that the latter does not involve any risk/return trade-off. That is, an investment with a high level of political risk does not offer a higher

---

[1] Pearlstein, S., "Understanding the Asian Economic Crisis," *The Washington Post*, January 18, 1998, p. A32.

potential return than one with a lower level of political risk. Thus, though an investor may seek out market risk as a means of earning a financial reward, higher levels of political risk offer no such promises of higher returns. One may experience the volatility without experiencing the reward. Another key difference is that the outcome of market risk is determined in the market, whereas the outcome of political risk is determined in the political arena. As a result, firms generally have much greater influence regarding market risks than regarding political risks. That is, a company can decide what to produce, where to produce, and other commercial decisions, but it generally has much less influence over political outcomes.

There are two types of political risk: sovereign risk and nonsovereign risk. *Sovereign risk* refers to actions taken by the government that affect commercial ventures. For example, in the years following World War II, socialist governments across Central and Eastern Europe expropriated private businesses. Other, less draconian sovereign risks include currency risks and risks involving changes in tax codes or legal statutes. Nonsovereign political risk refers to actions taken by nongovernmental organizations that have an impact on commercial activities. Terrorist groups that target businesses or kidnap corporate executives are a source of nonsovereign risk. Similarly, organized crime can present a type of nonsovereign political risk. In many parts of the former Soviet Union, for example, investors and entrepreneurs list organized crime as a significant impediment to investment.

### 11.2.2  Managing International Risk

International risk management dates to the time of the Greek city-states when shipping companies provided insurance against international losses.[2] In order for global commerce and trade to develop, merchants needed a means to protect themselves against catastrophic losses. The earliest form of political risk insurance compensated merchants in the event of attacks on their ships by pirates. As a result, international risk management developed in centers of trade and shipping such as Venice, Genoa, and more recently, London.

### 11.2.3  Diversification

As the global effects of the Southeast Asian crisis demonstrate, managing international risk is of critical importance to businesses and investors today. Just as the scale of international trade and commerce has increased tremendously in modern times, opportunities to manage risk have increased as well. Perhaps the simplest means of managing international political risk is through geographic diversification of investments. Much as a national investor might diversify her stock portfolio across industries to hedge against the risk of a single industry performing poorly, an international

---

[2]For an overview of the development of risk management and its importance in facilitating trade and commerce, see Haufler, V., *Dangerous Commerce: Insurance and the Management of International Risk*, Cornell University Press, Ithaca, NY, 1997.

investor can spread risks across different countries. A multinational company, for example, might build factories in disparate parts of the globe to hedge the risks in any one particular country. In order for such a strategy to be effective, the company must invest in countries whose political risks are not highly correlated. Thus, for risk in, say, Romania, an investment in Southeast Asia would provide a better hedge than an investment in another country in Southeast Europe.

A major drawback to this approach is that diversification is not always feasible. In some cases, investment is on such a massive scale that it is virtually impossible to offset the risks with another investment. In recent years, oil companies have made enormous investments in the Caspian Sea region in order to develop oil and natural gas resources they believe are located there. For example, Mobil, Exxon, Conoco, and Frontera Resources recently signed multibillion-dollar contracts for part ownership of Azerbaijani oil fields. According to the *Financial Times*, foreign oil producers intend to invest a total of $50 billion in Azerbaijan alone.[3] In addition to normal market risks, such as the difficulties involved in transporting oil and gas to Europe and the United States and the possibility that reserves are much smaller than expected, the oil companies are taking on significant political risk. Every one of the countries that border the Caspian Sea—Russia, Azerbaijan, Iran, Turkmenistan, and Kazakhstan—has experienced severe political upheavals in recent years. In this region, the political risks of large-scale foreign investment are substantial.

Since oil and gas reserves are located only in particular regions, opportunities for geographic diversification in this industry are extremely limited. Furthermore, the scale of the investment makes it difficult for the oil producers to reproduce such massive projects in other regions. As a result, these companies must seek alternative means to offset the considerable political risks of investment in the Caspian Sea region.

## 11.2.4 Political Risk and Export Credit Insurance

Another source of political risk management is straightforward political risk insurance. This type of insurance is available from both private and public institutions and provides for specific coverage against political risks. For example, Lloyd's of London offers policies that protect against, among other threats, confiscation and expropriation, political risk, and terrorism. AIG, another large insurance company, offers similar coverage against a myriad of political risks. Government agencies like the U.S. Overseas Private Investment Corporation (OPIC) also provide political risk insurance.

Export credit insurance provides another means of hedging international political risks. Insurance companies or government agencies provide guarantees to private banks that provide loans for foreign organizations to purchase domestic goods. The Export-Import Bank of the United States (Ex-Im Bank), for example, guarantees the

---

[3]Whalen, J., "Survey—World Energy: Western Groups Harbour High Expectations: Azerbaijan and Georgia," *Financial Times*, June 10, 1999.

repayment of loans for foreign purchasers of U.S. goods and services, and provides credit insurance against the risks of non-payment by foreign buyers for political or commercial reasons. In a typical transaction, the Ex-Im Bank guaranteed a loan made by Southtrust Bank to a hospital in West Africa that allowed the hospital to purchase $1 million of medical equipment from an American manufacturer.[4]

Though undoubtedly useful for hedging some kinds of international political risk, insurance is often prohibitively expensive. Furthermore, insurance is not always available in sufficient coverage amounts to cover some international investments. As a result, those involved in political risk management have looked to the massive derivatives market as a source for risk management tools.

## 11.3 CREDIT DERIVATIVES AND THE MANAGEMENT OF POLITICAL RISK

The market for derivatives has grown immensely since Black and Scholes published their influential paper on options pricing. Credit derivatives, designed to manage various forms of default risk, composed a market estimated at $200 billion by the mid-1990s. Though it is difficult to measure precisely, the market is growing rapidly and analysts expect that trading volume will reach $1 to $2 trillion by 2000. The array of products in the credit derivatives market is dizzying, and their applications are extremely diverse. This section describes several types of credit derivatives that are designed to hedge against international political risk.

### 11.3.1 Foreign Currency and Derivatives

For international investors, converting foreign currency into their own currency is a crucial issue. Derivatives allow investors to hedge against the risk that a currency will lose value. The most common type of derivative for hedging currency risks is a futures or forward contract that locks in an exchange rate at a particular date in the future. As you saw in Chapters 1 and 10, forwards and futures are deterministic contracts, and thus, their pricing is straightforward.

Aside from the actual exchange rate, there are also risks involved in converting foreign currencies. On numerous occasions, governments have, for a variety of political reasons, limited the ability of individuals to exchange the local currency for foreign currencies. As a result, international investors must, regardless of the actual rate of exchange, account for the possibility that they will not be able to convert the local currency. For example, a company that invests in Argentina may purchase a derivative contract that protects against the possibility that the Argentine government may not allow the exchange of currency. If that occurs during the period of the contract, the seller of protection, usually a bank or other financial institution, will pay the company the amount of money specified by the contract (and dependent on the nominal, albeit unobtainable, exchange rate).

---

[4]See the Ex-Im Bank web page for this and other examples: http://www.exim.gov/.

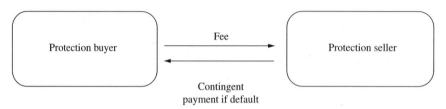

**FIGURE 11.1**
Default risk protection by credit derivative

## 11.3.2 Credit Default Risk and Derivatives

Whenever two parties enter into a contract, there is always the possibility that one or both will not meet all of the specified obligations. Credit default protection is designed to protect against this kind of risk. An investor with exposure to political risk could purchase a contract, usually from a bank or large financial institution, that provides a payment in the case of default by a foreign government. When insurance is not available or is considered too costly, investors will often turn to this type of derivative in order to hedge some or all of their political risk. The protection buyer pays a fee and receives a payment in the event of default of the reference asset.

Figure 11.1 shows the simplest case in which an investor hedges political risk by purchasing a credit derivative. In the event of a default on the reference asset, often a corporate or government bond in a foreign country, the protection buyer receives a payment from the protection seller. If there is no default, the protection seller keeps the original fee charged for the derivative and the protection buyer receives no payment.

An important strategy for hedging a foreign investment is to purchase a country risk derivative. This type of contract is based on the performance of a sovereign bond and provides the protection buyer with a payment in the event of default by a foreign government on its debt. As Figure 11.2 shows, the risk protection buyer pays a fee, usually a certain number of basis points, to a protection seller. If the

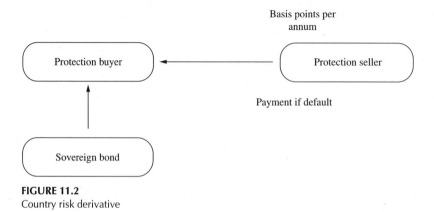

**FIGURE 11.2**
Country risk derivative

referenced foreign government defaults on its sovereign bonds, the protection buyer receives a predetermined payment. That provides a hedge against country risk, since if the government defaults on its debt, the investor can offset losses due to political risk with the gains from the derivative.

## A Specific Example of a Country Risk Derivative[5]

Several years ago, a U.S. high-tech company invested $100 million to build a factory in Mexico and sought to hedge the political risk associated with the investment. The company feared that such risks as war, labor strikes, confiscation or expropriation of its assets, or a coup d'état might disrupt its business. Insurance seemed too expensive, so the company decided to hedge its political risk with a credit default derivative. In addition to lower costs, the derivative offered the advantage of a quicker payment than an insurance contract.

The reference bond for this particular derivative was a U.S. dollar-denominated bond issued by the government of Mexico. The company paid 75 basis points for a year-long $100 million contract. In the event of a credit default event, the U.S. company would receive a payment equal to $100 million $\times$ (1 − market value of the bonds at the time of the occurrence of the credit event). Thus, if the sovereign bonds lost, say, half of their face value, the company would receive a payment of $50 million. The contract defined a credit event as:

> A bankruptcy event or payment default by the government, or any war, revolution, insurrection, or hostile act that interferes with foreign exchange transactions, or the expropriation, confiscation, requisition, or nationalizing of non-local banks, the declaration of a banking moratorium or suspension of payments by local banks.

Further, it specified payment default as the government failing "to pay any amount due of the referenced asset...of not less than USD $10 million" and bankruptcy event as "the declaration by the issuer of a general moratorium...amount of not less than USD $10 million." If a serious political or economic problem occurred in Mexico, the price of its government bonds would fall, and the buyer of protection would receive a payment based on the drop in bond prices.

Notice that the company's political risk was not perfectly correlated with the country risk derivative. Ideally, of course, in the case of a political catastrophe, management would like to have a payment exactly equal to the losses it suffers from political events. As discussed above, the company could have purchased an insurance policy that would provide a payment to offset exactly any loss in value of the factory in Mexico caused by political events. Due to the expense of insurance, and the time-consuming process of making a claim, the company decided that insurance was not the best option. Though not perfectly correlated with the company's losses in

---

[5]This example is drawn from Tavakoli, J. M., *Credit Derivatives: A Guide to Instruments and Applications*, New York: Wiley, 1998, pp. 84–87. Tavakoli's book does not discuss the mathematics of pricing credit default derivatives, but it provides an excellent overview of different products and many examples of their applications.

the event of a serious political problem, the credit derivative provided some payment to offset potential losses. Of course, the buyer of protection would attempt to ensure that the risk it faces is closely correlated with the underlying asset (in this example, the sovereign bond).

## 11.4  PRICING INTERNATIONAL POLITICAL RISK

The price of the country risk derivative discussed above was 75 basis points per year. In essence, that derivative placed a precise market value on the country risk inherent in investing in Mexico during the year of the contract. How do market participants determine the appropriate prices for country risk? The answer is that it is very difficult.[6] According to Keith Lewis (now with Bank of America), "classical risk-neutral pricing models are nearly worthless when it comes to credit derivatives." His view is echoed by Janet Tavakoli, who argues that "supply and demand drives this market." The most fundamental problem in pricing financial assets subject to default risk is determining the default probability. Regardless of whether the reference bond is a corporate bond or a sovereign bond, pricing a credit derivative requires determining, at least implicitly, the probability that default will occur. Tavakoli refers to this default probability as an "unknowable unknown."

The difficulty in pricing is enhanced by difficulties unique to emerging markets. Since emerging markets generally have the highest level of country risk, those difficulties are particularly relevant for pricing political risk. Compared to well-developed markets like those in the United States or Europe, emerging markets have comparatively higher volatility, fewer traders and less pricing data, fewer financial instruments, fewer pricing sources (e.g. Bloomberg's), and higher credit event risk. Furthermore, credit risks in emerging markets appear correlated, even in markets that are not closely linked. According to the International Monetary Fund,[7] for example, the crisis in Thailand in mid-1997 quickly spread with great force and persistence to Indonesia, Malaysia, the Philippines and, somewhat later, to Korea and, more briefly, to Hong Kong, Singapore, and Taiwan within the region, as well as to a number of emerging market economies in other regions.

---

[6]Country risk analysis ranges from highly technical to highly intuitive. The methods discussed in this chapter focus on mathematical models for valuing country risk. However, some managers rely on far more intuitive methods. An executive at a large U.S. pharmaceuticals company who was responsible for the firm's Latin American operations provided an example of the latter approach. He told me that on a trip to review the company's activities in a particular Latin American country, he arrived just after the election of a new government. The newly elected president's address reminded him of Fidel Castro's rhetoric that he had heard as a young man in Cuba in the early 1960s. Concerned with political risk under the new regime, he closed all of the company's operations in the country.

[7]"Financial Crises: Characteristics and Indicators of Vulnerability," *World Economic Outlook*, International Monetary Fund, May 1, 1998.

## 11.4.1   The Credit Spread
## or Risk Premium on Bonds

An important means of pricing country risk is to compare the yield on an emerging
market bond with that of a *riskless asset*, for which U.S. government debt usually
serves as a proxy. The **yield spread**, or difference in yields between a risky bond and
a comparable (in terms of the time until maturity) U.S. government bond, provides a
market measure of the amount of political risk in a particular country. For example,
on July 28, 1999, the yield spread between Mexican government debt and the U.S.
T-bill was 4.62%, or 462 basis points. Note that this particular bond is denominated
in U.S. dollars, so comparing the two assets does not require considering the dollar-
peso exchange rate.[8]

Table 11.1 gives examples of the yield spread for other dollar-denominated
debt issued by emerging market governments. For each sovereign bond listed in the
table, the yield spread gives a measure of how much more risk holding the emerging market bond carries vis-à-vis a *riskless* U.S. government bond.[9] Implicit in this
yield spread is a market indication of the probability that the sovereign borrower

TABLE 11.1
**Credit ratings and yield spreads
for emerging government debt**

| Sovereign Bond | Redemption Date | S&P Rating | Yield Spread vs. U.S. |
|---|---|---|---|
| Croatia | 03/2006 | BBB− | 2.55 |
| Slovenia | 03/2009 | A | −0.18 |
| Hungary | 02/2009 | BBB | −0.07 |
| Argentina | 09/2027 | BB | 7.47 |
| Brazil | 05/2027 | B+ | 8.27 |
| Mexico | 05/2026 | BB | 4.62 |
| China | 12/2008 | BBB+ | 2.23 |
| Philippines | 01/2019 | BB+ | 4.27 |
| South Korea | 04/2008 | BBB− | 2.57 |
| Lebanon | 07/2000 | BB− | 1.84 |
| South Africa | 04/2008 | BB+ | 3.04 |
| Turkey | 09/2007 | B | 5.53 |

---

[8]Even though the bonds are dollar-denominated, this does not mean that the exchange rate is irrelevant.
The current exchange rate may have a strong bearing on determining whether or not the government
can meet its obligations. For example, if the currency of an emerging market collapses, the government
will have more difficulty paying its dollar-denominated debt, since it will need much more of the local
currency. Thus, although the exchange rate is not explicitly a factor in comparing U.S. T-bonds to dollar-
denominated emerging market bonds, it is an important implicit factor.

[9]Note that country risk is not the only determinant of the yield spread. Tax considerations, capital con-
trols, home-country bias, local market conditions, and other factors can affect the supply and demand
for emerging market bonds.

will default on its bonds, the "unknowable unknown" described by Tavakoli. Thus, a model that could accurately determine the yield spread would contain information about the credit risk inherent in investing in that bond. This is the key piece of information needed to price the type of country risk derivatives discussed above. Recall that the U.S. high-tech company paid 75 basis points to hedge its $100 million investment in Mexico. This price accounts for the extra risk inherent in Mexican bonds for the duration of the derivative contract.

## 11.5 TWO MODELS FOR DETERMINING THE RISK PREMIUM

For simplicity, the bond pricing models discussed in Chapter 8 assume that the bonds are riskless. Even for U.S. government debt, considered among the least risky debt in the world, this assumption is not entirely accurate. For emerging market debt, the assumption of no default risk is clearly not applicable. In order to deal with this problem, analysts have developed mathematical models designed to price risky debt. One class of models is based on Black and Scholes's seminal paper on options pricing. This approach treats the probability of default as an endogenous variable. A second approach takes the credit default probability as given based on credit ratings and then uses this information to calculate the yield spread. The next two sections discuss these two broad approaches to pricing risky debt.

### 11.5.1 The Black-Scholes Approach to Pricing Risky Debt

The Nobel Prize-winning economist Robert Merton applied the Black-Scholes method for pricing financial instruments to the case of corporate debt.[10] Merton's model provides a means for calculating the yield spread between risky bonds and riskless bonds. The first step is to describe the dynamics of the value of a firm:

$$dV = (\mu V - C)dt + \sigma V dB \qquad (11.1)$$

where

$\mu$ = expected rate of return on the firm
$C$ = payouts by the firm per unit of time to its shareholders
$\sigma^2$ = variance of the return on the firm per unit of time
$B_t$ = a standard Brownian motion

Thus, changes in the value of the firm are explained by the rate of return on the firm minus its payoffs to shareholders, plus a term that captures random changes in its value.

---

[10]Merton, R. C. "On the Pricing of Corporate Debt: The Risk Structure of Interest Rates," *Journal of Finance*, 29, 2 (May 1974), pp. 449–470.

Suppose we have a security whose market value, $Y$, can be written as a function of the value of the firm, $V$, and the time:

$$Y = F(V, t) \tag{11.2}$$

For clarity in discussing this security, we first present the case where $C = 0$ (no payout by the firm), and we also assume that the security pays no dividends.

You will recognize that, in this case, $Y$ may be modeled as a derivative security based on the underlying value of the firm, $V$. Also, equation (11.1) with $C = 0$ is our familiar model from Chapter 6 for geometric Brownian motion. We may use the Black-Scholes argument of Chapter 6 to obtain the following PDE for $Y = F(V, t)$:

$$\tfrac{1}{2}\sigma^2 V^2 F_{vv} + rV F_v + F_t = rF \tag{11.3}$$

Equation (11.3) is a partial differential equation that must be satisfied by any security whose value is a function of the value of the firm and time. By using different boundary conditions, it is possible to describe different types of securities (e.g., equity vs. bonds). In this section, we will use specific boundary conditions in order to generate the spread between risky and riskless bonds. Note that in addition to the value of the firm and time, $F$ depends on the interest rate, $r$, and the volatility of a firm's value, $\sigma^2$.

### Zero-Coupon Bonds

We assume that the value of a ZCB issued by the firm can be expressed as $F(V, t)$. Then, using (11.3), we have the following PDE for the bond price:

$$\tfrac{1}{2}\sigma^2 V^2 F_{vv} + rV F_v - rF - F_\tau = 0 \tag{11.4}$$

where $\tau \equiv T - t$ is the length of time until maturity, so that $F_t = -F_\tau$. To solve (11.4), two boundary conditions and an initial condition are necessary. Those conditions are derived from the structure of the bond issue.

By definition, the value of the firm is equal to the value of its equity plus the value of bond issues ($V = F(V, \tau) + f(V, \tau)$, where $f$ is the value of the firm's equity). According to U.S. laws concerning limited liability, the value of the equity and the value of the bonds cannot be negative.[11] If the value of the firm is zero, then the value of the equity and bonds will also be zero. Hence,

$$F(0, \tau) = f(0, \tau) = 0 \tag{11.5}$$

Further, the value of the security cannot exceed the value of the firm ($F(V, \tau) \leq V$), which implies that

$$\frac{F(V, \tau)}{V} \leq 1 \tag{11.6}$$

---

[11] *Limited liability* means that an investor who buys shares in a company is only liable for the amount of money that he or she invested. Suppose that a company lost a court case and was ordered to pay damages that far exceeded its assets. The company would go bankrupt, but its shareholders would not become liable for the debts of the company. Once the value of their shares falls to zero (because of the firm's bankruptcy), the shareholders are not subject to further losses.

The initial condition is derived from the fact that the firm will pay the amount $B$ to bearers of its debt at maturity. If, at maturity, the value of the firm, $V$, is less than the payoff, $B$, the firm will go into default and the bondholders will receive $V$. If the value of the firm exceeds the value of the bond payment, the firm will pay $B$. Thus, at maturity, the minimum value of the bonds will be the lesser of the value of the firm and the value of the payment. Thus, the initial condition is

$$F(V, 0) = \min\{ V, B \} \tag{11.7}$$

With the three conditions (11.12), (11.13), and (11.14), it is possible to solve (11.4) directly for the value of the debt. Merton, however, arrives at the solution by examining a closely related problem that already has a known solution.

To determine the value of equity, $f(V, \tau)$, we note that $f = V - F$ and substitute for $F$ in (11.4). The result is a PDE for $f$:

$$\tfrac{1}{2}\sigma^2 V^2 f_{vv} + rV f_v - rf - f_\tau = 0 \tag{11.8}$$

subject to

$$f(V, \tau) = \max\{ 0, V - B \} \tag{11.9}$$

and the two boundary conditions (11.5) and (11.6). The initial condition specifies that the value of equity when a firm's debt reaches maturity is equal to the value of the firm minus the payoff to the debt holders. If the value of the firm is less than the payoff, then the equity is worthless. This initial condition is rooted in U.S. law, which holds that a firm is obligated to pay the holders of its debt before it pays the holders of its equity. Thus, the shareholders receive $V - B$ unless this is less than zero. In that case, since shareholders are subject only to limited liability, they receive nothing but have no further obligation to pay the bondholders.

Notice that if you replace the firm value, $V$, and the bond payoff, $B$, equations (11.8) and (11.9) are identical to the equations for pricing a European call option in Chapter 6 with the stock price $S$ and the exercise price $X$, in place of $V$ and $B$ respectively. Thus, it is possible to apply the solution for the call option to the case of risky debt. When $\sigma^2$ is a constant,

$$f(V, \tau) = VN(d_1) - Be^{-r\tau}N(d_2) \tag{11.10}$$

where

$$d_1 \equiv \frac{\left\{ \ln\left[\dfrac{V}{B}\right] + \left(r + \dfrac{1}{2}\sigma^2\right)\tau \right\}}{\sigma \sqrt{\tau}}$$

and

$$d_2 \equiv d_1 - \sigma \sqrt{\tau}$$

From equation (11.10) and the fact that $F = V - f$, we can write the value of the debt issue as

$$F(V, \tau) = V(1 - N(d_1)) + Be^{-r\tau}N(d_2)$$

so that

$$F(V, \tau) = B e^{-r\tau} \left\{ N(d_2) + \frac{V}{B} e^{r\tau} N(-d_1) \right\}. \tag{11.11}$$

Equation (11.11) provides a means for pricing risky debt. However, in order to calculate the yield spread, which provides a measure of the risk of a bond issue, we can rewrite (11.11) as

$$F = B \exp[-R(\tau)\tau]$$

That is, we are defining the yield of this risky bond as

$$R(\tau) = -r - \frac{1}{\tau} \ln \left\{ N(d_2) + \frac{V}{B} e^{r\tau} N(-d_1) \right\} \tag{11.12}$$

Note that the risk premium is a function of only two variables:

1. The variance (or volatility) of the firm's operations, $\sigma^2$
2. The ratio of the present value (at the riskless rate) of the promised payoff to the current value of the firm; that is, $V/Be^{-r\tau}$

### Calculating the Risk Premium Using the Black-Scholes Method

Merton's adaptation of the Black-Scholes model is designed to price corporate debt. Thus, if we want to find the yield spread (i.e., the risk premium) on a corporate bond, we need to calculate $\sigma^2$ and $d$. Suppose that a company worth \$1,000,000 has \$500,000 in outstanding bonds that will mature in one year. In order to determine the risk premium on those bonds, we need to know the variance, $\sigma^2$, and the current interest rate, $r$. Suppose that they are 0.25 and 0.05 respectively. Thus, we have

$$V = 1,000,000$$
$$B = 500,000$$
$$\sigma^2 = 0.25$$
$$\tau = 1$$

Plugging these values into (11.12) provides the following result:

$$d_1 = 1.236294361$$
$$d_2 = 2.236294361$$
$$N(d_1) = 0.04125586$$
$$N(d_2) = 0.891825352$$
$$R(\tau) = 0.021665473$$

Thus, the bonds will trade at a premium of 2.17% over the riskless rate of 5%.

Merton's model is designed to calculate the risk premium on *corporate* bonds. However, the model is applicable to government bonds, and thus it allows us to calculate the risk premium on sovereign debt. Recall that in order to find the risk premium, the model requires the value of a firm and the volatility of that value. Like a company, governments have a balance sheet of assets and liabilities and the performance of that balance sheet varies over time, so the "value" of the government fluctuates. Like the U.S. government, many foreign governments borrow money and are able to continue borrowing for long periods of time. In addition, government can exact resources from their citizens in the form of taxes. However, the amount that government can borrow is limited, as is the amount of money that governments can extract from their citizens. Furthermore, there is some minimal level of services that a government must provide. Thus, if debt payment become too great, the government will not be able to borrow, tax, or cut spending in sufficient quantities to meet its obligations. As a result, governments can, and occasionally do, default on their debts.

For example, in 1998 the Russian government defaulted on a part of its debt, because it could no longer make the required interest payments. The model described above can thus capture the risk premium on sovereign bonds and hence provide some means of measuring country risk. Suppose that the government of a small country is above its default level by, say, $10 billion. Further, assume that figure includes $8 billion in outstanding debt and that the volatility of the government's value is 0.75. Thus, we have

$$V = 10,000,000,000$$
$$B = 8,000,000,000$$
$$\sigma^2 = 0.75$$
$$\tau = 0.5$$

According to the model, the risk premium for this country's debt is 33%, which is a very high spread. The reason is that the country is quite close to the limit where its debt is not sustainable. Further, there is a high level of volatility, which suggests that the government may cross that limit and thus go into default. Thus, Merton's analysis is directly applicable to sovereign bond spreads, provided that the value of the government balance sheet and the volatility of that value can be found.

### 11.5.2 An Alternative Approach to Pricing Risky Debt

Some authors have pointed out the difficulty of determining the exact value of a firm, which is necessary for implementing Merton's model. Further, critics have noted that determining the precise structure and amount of all liabilities is difficult. As a result, analysts have developed models that attempt to avoid those purported problems. One proposed solution is to utilize the information contained in credit ratings in order to price risky debt. In a 1997 article, Jarrow, Lando, and Turnbull (JLT) describe a model[12] that prices risky debt by assuming that the probability of default is exogenous. The model relies on credit ratings and historical patterns of default to generate

---

[12]Jarrow, R. A., Lando, D., and Turnbull, S. M., "A Markov Model for the Term Structure of Credit Risk Spreads," *Review of Financial Studies*, 10, 2, (Summer 1997), pp. 481–523.

the default probabilities needed to price risky bonds. The model first describes the price of a riskless bond. Assuming arbitrage-free and complete markets, the default-free bond price is the expected, discounted value of a sure dollar received at time $T$:

$$P(t, T) = \tilde{E}\left(\frac{B(t)}{B(T)}\right) \tag{11.13}$$

where

$$B(t) = \exp\left(\int_0^t r(s)ds\right)$$

Equation (11.13) states that the price of a bond today is found by discounting the future payoff at the interest rate $r$.

The price of a bond subject to default risk, then, is the expected value of a *risky* dollar received at time $T$:

$$v(t, T) = \tilde{E}\left(\frac{B(t)}{B(T)}(\delta 1_{(\tau^* \le T)} + 1_{(\tau^* > T)})\right) \tag{11.14}$$

where

$v(t, T)$ = the price of a risky zero-coupon bond

$\delta$ = the recovery rate or the fraction of each dollar received in the event of a default

$\tau^*$ = the random time at which bankruptcy occurs

$1_{(x \le a)}$ = 1 for $x \le a$ and 0 for $x > a$

Equation (11.14) shows that if there is no default, the value of the bond is exactly the same as in the riskless case. If the issuer defaults, though, the bearer of the bond will receive some portion, $\delta$ (delta), of the riskless value. Delta is called the **recovery rate** because even in cases of bankruptcy and default, bond-holders usually receive at least some portion of the money owed to them.

Assuming that the stochastic process for default-free spot rates and the bankruptcy process are statistically independent, the price of a zero-coupon risky bond may be written as

$$v(t, T) = \tilde{E}_t\left(\frac{B(t)}{B(T)}\right)\tilde{E}_t[\delta 1_{(\tau^* \le T)} + 1_{(\tau^* > T)}] \tag{11.15}$$

$$= P(t, T)[\delta + (1 - \delta)\tilde{Q}_t(\tau^* > T)]$$

where

$\tilde{Q}_t(\tau^* > T)$ is the probability under $\tilde{Q}$ that default occurs after date $T$

Thus, the price of a risky bond today is simply a weighted average of the payoff in the event of default and the payoff if no default occurs. Specifically, the second factor in equation (11.15) is the sum of the probability of default $(1 - Q)$ multiplied by delta, plus the probability of no default $(Q)$ multiplied by one.

$Q$ is a transition matrix, in which each entry gives the probability of moving from one asset class to another:

$$Q_{t,t+1} = \begin{bmatrix} q_{11}(t,t+1) & q_{12}(t,t+1) & \cdots & q_{1K}(t,t+1) \\ q_{21}(t,t+1) & q_{22}(t,t+1) & \cdots & q_{2k}(t,t+1) \\ \vdots & \vdots & \ddots & \vdots \\ q_{K1}(t,t+1) & q_{K2}(t,t+1) & \cdots & q_{KK}(t,t+1) \end{bmatrix} \quad (11.16)$$

Thus $q_{ij} \equiv$ the probability of moving from asset class $i$ to asset class $j$.

The transition matrix is based on credit ratings provided by companies like Standard & Poor's or Moody's. Table 11.2 provides average one-year transition probabilities for corporate bonds. For example, a bond with a BBB rating has a 0.0006 probability of ending the year with a rating of AAA. The same bond has a 0.7968 chance of ending the year with the same rating and a 0.0043 chance of going into default (a D rating). Perhaps not surprisingly, for each bond rating, the probability of remaining in that category (the diagonal of the matrix) is the highest. Note also that there is no row corresponding to D, because if an issuer goes into default, there is no possibility that it will change categories.

In chapters 5 and 6 we used no-arbitrage probabilities in order to price options. The JLT model relies on the same technique, which implies that historical default probabilities are not sufficient to price risky bonds. In other words, a credit rating organization like Standard & Poor's develops the historical default probabilities based on observed rates of default. Those observed default rates include the risk premiums that investors require in order to hold risky debt. The JLT model, on the other hand, requires risk-neutral probabilities. Thus, the model adjusts the historical default probabilities in order to include risk premiums for holders of risky bonds. The transition matrix under the risk-neutral probability is given by

$$\widetilde{Q}_{t,t+1} = \begin{bmatrix} \widetilde{q}_{11}(t,t+1) & \widetilde{q}_{12}(t,t+1) & \cdots & \widetilde{q}_{1K}(t,t+1) \\ \widetilde{q}_{21}(t,t+1) & \widetilde{q}_{22}(t,t+1) & \cdots & \widetilde{q}_{2K}(t,t+1) \\ \vdots & \vdots & \ddots & \vdots \\ \widetilde{q}_{K1}(t,t+1) & \widetilde{q}_{K2}(t,t+1) & \cdots & \widetilde{q}_{KK}(t,t+1) \end{bmatrix} \quad (11.17)$$

**TABLE 11.2**
**One-year transition probabilities for corporate credit ratings***

| Initial Rating | End-of-year Rating | | | | | | | | |
|---|---|---|---|---|---|---|---|---|---|
| | AAA | AA | A | BBB | BB | B | CCC | D | NR |
| AAA | 0.8746 | 0.0945 | 0.0077 | 0.0019 | 0.0029 | 0.0000 | 0.0000 | 0.0000 | 0.0183 |
| AA | 0.0084 | 0.8787 | 0.0729 | 0.0097 | 0.0028 | 0.0028 | 0.0000 | 0.0000 | 0.0246 |
| A | 0.0009 | 0.0282 | 0.8605 | 0.0628 | 0.0098 | 0.0044 | 0.0000 | 0.0009 | 0.0324 |
| BBB | 0.0006 | 0.0041 | 0.0620 | 0.7968 | 0.0609 | 0.0151 | 0.0043 | 0.0043 | 0.0545 |
| BB | 0.0004 | 0.0020 | 0.0071 | 0.0649 | 0.7012 | 0.0942 | 0.0218 | 0.0218 | 0.0970 |
| B | 0.0000 | 0.0017 | 0.0027 | 0.0058 | 0.0451 | 0.7196 | 0.0598 | 0.0598 | 0.1272 |
| CCC | 0.0000 | 0.0000 | 0.0102 | 0.0102 | 0.0179 | 0.0665 | 0.2046 | 0.0246 | 0.1176 |

*Source: Tavakoli, J. M., *Credit Derivatives: A Guide to Instruments and Applications.* Wiley, New York, 1998, p. 117.

where

$$\widetilde{q}_{ij} \equiv \text{ the probability of moving from one asset class to another}$$

and

$$\widetilde{q}_{ii}(t, t + 1) \equiv -\sum_{j \neq i}^{K} \widetilde{q}_{ij}(t, t + 1) \tag{11.18}$$

Each of the entries in the matrix comprises a transition probability and a risk adjustment, so (11.16)

$$\widetilde{q}_{ij}(t, t + 1) = \pi_i(t) q_{ij} \tag{11.19}$$

Equation (11.19) says that the risk-neutral probability associated with a change in credit ratings is composed of the empirical probability multiplied by a risk premium. In the examples in this chapter, $\pi$ is always greater than or equal to 1, so the practical effect is to increase the probability that an asset changes credit classes and decrease the probability that an asset remains in the initial credit class. The economic interpretation is that by using the risk-neutral probabilities (and hence increasing the probability of default), the JLT model produces a lower price and higher yield for the bond than if the empirical probabilities were used. In other words, if one used the risk-neutral pricing method employed by the JLT model but empirical default probabilities that included risk premiums, the prices generated by the model would be artificially high and the credit spread artificially low. That is because a risk-neutral investor would be willing to tolerate more risk without extra compensation vis-à-vis real-world, risk-averse investors. Thus, the model takes the risk-average default probabilities and increases the probability of default, thus generating the lower prices and higher yields.

Let us look at a simple example of the difference between empirical and no-arbitrage default probabilities. Suppose that there are only three credit classes, A, B, and D (for default), and the empirical transition matrix looks like

**End-of-year credit rating**

| Initial credit rating | A | B | D |
|---|---|---|---|
| A | 0.8 | 0.1 | 0.1 |
| B | 0.1 | 0.8 | 0.1 |
| D | 0 | 0 | 0 |

If $\pi_i = 1.5$, then using equations (11.17) and (11.19), the risk-neutral transition matrix is

**End-of-year credit rating**

| Initial credit rating | A | B | D |
|---|---|---|---|
| A | 0.7 | 0.15 | 0.15 |
| B | 0.15 | 0.7 | 0.15 |
| D | 0 | 0 | 0 |

As described above, the practical effect is to increase the probability of default and decrease the probability of remaining in the same credit class.

With this structure, it is possible to write an expression for the credit risk spread between a no-risk asset and a risky asset. The price of a zero-coupon bond issued by a borrower in credit class $i$ is given by

$$v^i(t, T) = p(t, T)(\delta + (1 - \delta)\widetilde{Q}^i_t(\tau^* > T) \qquad (11.20)$$

The forward rate for the risky zero-coupon bond in credit class $i$ is defined as

$$f^i(t, T) = -\ln\left(\frac{v^i(t, T + 1)}{v^i(t, T)}\right) \qquad (11.21)$$

Substitution yields

$$f^i(t, T) = f(t, T) + 1_{(\tau^* > 1)} \ln \frac{[\delta + (1 - \delta)\widetilde{Q}^i_t(t^* > T)]}{[\delta + (1 - \delta)\widetilde{Q}^i_t(t^* > T + 1)]} \qquad (11.22)$$

$$f = \text{forward rate for a riskless asset}$$

$$f^i = \text{forward rate for a risky asset}$$

This equation provides an expression for the credit spread, or risk premium, in which the difference between the forward rate on a risky asset and a riskless asset is a function of the probability of default and the recovery rate. To get the spot rate, set $T = t$ and simplify:

$$r^i(t) = r(t) + \ln \frac{1}{[1 - (1 - \delta)\widetilde{q}_{ik}(t, t + 1)]} \qquad (11.23)$$

Thus, estimates of the recovery rate and the risk-neutral transition probability are necessary to generate the theoretical credit risk spread. Based on historical data, one can compute an estimate of the recovery rate. As discussed above, combining the empirical transition matrix with the required risk premium, $\pi$, provides an estimate of the risk-neutral transition probability matrix. Using some algebra, the JLT article shows that $\pi_i$ can be found using the following equation:

$$\pi_i(t) = \sum_{j=1}^{K} \widetilde{q}^{-1}_{ij}(0, t) \frac{p(0, t + 1) - v^i(0, t + 1)}{p[0, t + 1(1 - \delta)q_{iK}]} \qquad (11.24)$$

With this equation for $\pi_i$ and the empirical transition matrix, it is possible to calculate the theoretical credit spread for bonds subject to credit risk.

## 11.6   A HYPOTHETICAL EXAMPLE OF THE JLT MODEL

Suppose that you want to predict the spread on a sovereign bond issued by the government of Croatia with maturity in March 2006. You find the following information and use that with equation (11.24):

$$\text{Recovery rate} = \delta = 0.2$$
$$\text{Default probability} = \widetilde{q}_{iK}(t, t + 1) = 0.0085 \qquad (11.25)$$

This yields

$$\text{Yield spread} = r^i(t) - r(t) = \ln\frac{1}{1 - (0.8 \times 0.0295)} = 0.0239 \quad (11.26)$$

Thus, the model predicts a difference of 239 basis points, or 2.39%, between the interest rate on U.S. government bonds and Croatian government bonds of the same maturity.

## EXERCISES

1. Explain in economic terms why the assumption that $e^{[-R(\tau)\tau]} \equiv F(V, \tau)/V$ in Merton's model is reasonable. Show how to derive equation (11.11) from (11.10).
2. Derive equation (11.13) in Merton's model from equation (11.12).
3. Explain in economic terms why $p(t, T) = E[B(t)/B(T)]$.
4. Derive equation (11.24) from the JLT model.
5. Find the following risk premiums using the Black-Scholes method:

|   | V | B | r | $\tau$ | $\sigma^2$ |
|---|---|---|---|---|---|
| A | $10 million | $5 million | 0.10 | 0.5 | 0.2 |
| B | $100 million | $75 million | 0.05 | 0.5 | 0.5 |
| C | $100 million | $75 million | 0.05 | 0.5 | 0.75 |
| D | $1 billion | $800 million | 0.10 | 1 | 0.5 |
| E | $1 billion | $800 million | 0.10 | 2 | 0.5 |

6. Explain in economic terms why the risk premium in C is greater than the risk premium in B. Why is it less in E than in D?
7. Find the following risk premiums using the JLT method:

|   | $\delta$ | $\tilde{q}_{iK}(t, t+1)$ |
|---|---|---|
| A | 0.2 | 0.01 |
| B | 0.2 | 0.02 |
| C | 0.5 | 0.02 |

8. Explain in economic terms why the risk premium in A is less than the risk premium in B. Why is the risk premium in C greater than the risk premium in B?

# ANSWERS TO SELECTED EXERCISES

## Section 2.2

**1.** $a = 0.33, V = 4.64$

**3.** $a = -0.125, V = 2.96$

## Section 2.4

**1.** $a = -1, V = 15.7$

**2.** $a = -0.4, V = 2.89$

**5.** $V = \exp(-rt)(E[S_1] - X)$

## Section 2.5

**1.** (*a*) $V = \$2.75$; (*b*) 250; (*c*) $100

**3.** (*a*) $V = \$5.68, 1333.3, \$200$; (*b*) $V = \$1.94, -250, \$100$; (*c*) $V = \$18.70, 1666.6, \$200$; (*d*) $V = \$2.59, -4000, \$600$; (*e*) $V = \$2.47, 2000, \$400$; (*f*) $V = \$0.91, -600, \$300$

## Section 2.7 Review

**1** (*a*) 7.84; (*b*) 2.81; (*c*) 1.73; (*d*) 25.03; (*e*) 6.61; (*f*) 9.43

**3.** Sell many, and hedge to cover the liability

**5.** You hedge by buying 250,000 shares; your profit is $100,000.

## Section 3.1.2

**1.** $2.66; the two Tues. nodes are $2.18 and $3.63.

**3.** $pu + (1 - p)d = 1$

## Section 3.2

**1.** $39.9

**3.** $8.81

## Section 3.3

**1.** (*a*) 14.79; (*b*) 7.40; (*c*) 12.26; (*d*) 11.54; (*e*) 9.04; (*f*) 42.02; (*g*) 29.40

**3.** (*a*) 1.16; (*b*) 20.00; (*c*) 15.59; (*d*) 1.67; (*e*) 2.96; (*f*) 26.11; (*g*) 3.97

## Section 3.4

**1.** 23.14

**3.** 5.29

## Section 3.5

**1.** Using $r = 0$, $V = 109.33$

**3.** Using $r = 0$, $V = 58.64$

## Section 3.7

**1.** (c) For the path *uuuu*, the delta values are .57, .69, .86, 1.0. For the path *udud*, the delta values are .57, .69, .43, .64. For the path *dudu*, the delta values are .57, .30, .43, 0.0.

**3.** (e) For the path *uuuu*, the delta values are $-.74$, $-.17$, .0, .0. For the path *dddd*, the delta values are $-.74$, $-.93$, $-1$, $-1$, for the path *dudu* the delta values are $-.74$, $-.93$, $-.75$, $-1$.

## Section 4.1

**3.** Since $d = 1/u$, an up move, if followed by a down move, is canceled.

**4.** Figure 4.1 is based on the relation $ud = .99$. An up move followed by a down move results in a slight drop in value.

## Section 4.2

**1.** $V = 16.43$, 20.26, 24.23

**3.** $V = 29.36$

## Section 4.3

**1.** $V = 4.32$, no early exercise

**3.** $V = 1.47$, early exercise

## Section 4.4

**1.** $V = .31$

**3.** Option #1 = 23.27; Option #2 = 19.10

## Section 5.3

**1.** Monthly values $\mu = 0.0867$, $\sigma = 0.1117$; weekly values $\mu = 0.02$, $\sigma = 0.0537$; daily values $\mu = 0.0028$, $\sigma = 0.0203$. Sigma becomes larger relative to mu.

## Section 5.4

**1.** (*a*) 11.84; (*b*) 1.95; (*c*) 2.37; (*d*) 12.30; (*e*) 4.92; (*f*) 14.45; (*g*) 2.31; (*h*) 1.55

**2.** (*a*) 0.98; (*b*) 7.29; (*c*) 10.02; (*d*) 10.48; (*e*) 11.14; (*f*) 10.79; (*g*) 0.16; (*h*) 1.24

## Section 5.6

**1.** (*a*) Use equation (5.5) and choose $c = \sigma \cdot T$.

**3.** $\ln(S/X) + rT = \ln(S/X) + \ln(\exp(rT)) = \ln(\exp(rT)S/X)$

## Section 5.7

**1.** Model X with $2.7386Z + 30$: $\Pr[Z > 3.6515] = 0.001$

**3.** $u = 1.0166$, $d = 0.9837$, $q = 0.4979$

**5.** Use $n = 50$ steps: $u = 1.036$, $d = 0.9653$, $q = .4951$

## Section 6.4

**1.** $\partial V/\partial S = a$, $\partial V/\partial t = rbe^{rt} = rV - arS$; so $a = 1$ corresponds with the price of a forward contract.

**3.** $\partial V/\partial t = rV + e^{rt}\partial G/\partial t$; substitute this into the Black–Scholes equation.

## Section 6.6

**3.** $\partial G/\partial t = rG + e^{-rT+rt}\partial V/\partial t$; substitute this into equation (6.18).

## Section 7.3

**1.** (*a*) 0.45; (*b*) 0.25; (*c*) 0.195; (*d*) 0.25; (*e*) 1.05; (*f*) 0.048; (*g*) 0.673

## Section 7.4

**1.** Delta values: $(a)$ 0.411; $(b)$ 0.564; $(c)$ 0.78; $(d)$ 0.216; $(e)$ 0.973; $(f)$ 0.243

|     | Delta  | Gamma  | Theta   |
|-----|--------|--------|---------|
| $(a)$ | 0.4240 | 0.0003 | −0.0137 |
| $(b)$ | 0.5852 | 0.0004 | −0.0255 |
| $(c)$ | 0.7951 | 0.0002 | −0.0304 |
| $(d)$ | 0.2233 | 0.0003 | −0.0097 |
| $(e)$ | 0.9793 | 0.0002 | −0.0048 |
| $(f)$ | 0.2533 | 0.0007 | −0.0057 |

## Section 8.3.2

**1.** $-\ln P = R(T - t)$, and so $-\partial \ln P / \partial T = R$

**3.** ZCB prices: $(a)$ 0.9465; $(b)$ 0.8869; $(c)$ 0.6873; $(d)$ 0.3679; $(e)$ 0.0025

## Section 8.3.4

**1.** 0.9867

**3.** 0.7927

**5.** 0.2214

**7.** 0.5902

**9.** The annualized rate is 5.48%. $P(0, 1.5) = \exp(-1.5(.0548)) = 0.9211$

**11.** 0.9749

## Section 8.4.2

**1.** $0.0242/4.93055 = 0.00491$

**3.** $.2716/38.122075 = 0.0071$

## Section 8.5.4

**1.** $(a)$ 1.1026; $(b)$ $1/1.1026$; $(c)$ $1/1.05$; $(d)$ 1.05; $(e)$ Note first that the "market" interest rate for the period [0, 1] is 1.0509524 and second that $1.1026 = (1.05) \cdot (1.0509524)$.

**2.** 0.907112

## Section 8.8

**1.** $\sigma(s, t)^2 - \sigma(s, T)^2 = (t - T)(t + T - 2s)$; integrate this expression from 0 to $t$. Equation (8.49) can then be solved:

$$\ln P(t) = \ln P(0, T) - \ln P(0, t) + .5(t - T)tT + (T - t)B(t)$$

Now use equation (8.50)—differentiate this expression to solve for $r$.

**3.** $\sigma(s, t)^2 - \sigma(s, T)^2 = s^2(t - T)(t + T - 2s)$; integrate this expression from 0 to $t$; equation (8.49) can then be solved; again, use equation (8.50)—differentiate this expression to solve for $r$.

## Section 9.1

**1.** Using $\sigma = 0.011$, the first branch has $r = 0.05$; $0.333\sigma = 0.0064$. The next two branches will have $r$ values of $0.0515 + 0.0064$ and $0.0515 - 0.0064$.

## Section 10.1

**1.** (*a*) American to European: one mark = 0.4662 dollars; one franc = 0.139 dollars; one kroner = 0.1222 dollars; one pound = 1.5065 dollars; one peseta = 0.0055 dollars; one yen = 0.0091 dollars. (*b*) European to American: one dollar = 44.2478 francs; one dollar = 1.7068 francs; one dollar = 9.0662 krona; one dollar = 2.4172 guilder; one dollar = 1.8235 real; one dollar = 1.5567 C. dollars.

## Section 10.2

**1.** Forward exchange rates, 0.4678

**3.** 0.0092

## Section 11.6

**1.** Why is $F = e^{-Rt}V$ reasonable? $V$ represents the value of a firm, while $F$ denotes a security value determined by the market. The uncertainty regarding value and future value would result in some type of discounting—from value to market value. This defines the $R$ value.

**3.** Why is $P(t, T)$ an expected value of $B(t)/B(T)$? The payoff to a holder of this bond will certainly be $1. But this will occur at a future date, so its present value is obtained by discounting. The factor $B(t)/B(T)$ is the (uncertain) discount factor. The present value of the bond is the expected value of this quantity.

# INDEX

## A

arbitrage, 2, 5, 25, 27, 32, 50, 66, 69, 75, 76, 150
Asian financial crisis, 222
average relative return, 62

## B

back induction, 161, 164
Bernoulli random variable, 62, 63
binomial model, 66, 194
binomial probability, 98, 99
binomial random variable, 99, 194
binomial short rate, 195
binomial tree, 25, 39, 44, 192, 194
binomial Vasicek model, 200
Black-Scholes, 81, 225, 230
    equation, 113–115, 123, 130, 173, 212
    formula, 81, 90–92, 98, 103, 134
bond
    calibration, 176, 179, 181, 183
    risk premium, 223, 229, 230, 233
bond dynamics, 180–182
bond models, 137, 157, 171, 176, 178, 184, 192, 230
    simple, 174
    Vasicek, 178

bond price, 138, 140, 149, 162, 182, 199, 230
bond price differential, 182
bond price PDE, 173, 231
bond volatility, 181
Brownian motion, 81, 88, 106–108
Brownian paths, 106

## C

calibration
    bond model, 176, 185
    GBM, 87
    pricing model, 93
    spreadsheet, 89, 90
    tree, 61–64
call, 25
    American, 8
    European, 8
    future, 118
central limit theorem, 99
chaining, 46–48, 50, 56
chaining method, 53
collateral, 159
counterparty, 209
country risk, 222
    example, 227
coupon bonds, 16

coupon rate, 16
credit default risk, 226
credit rating, 235, 236, 237
credit spread, 229, 238
currency
    American convention,
      207
    European convention,
      207
currency forwards, 209
currency market, 207
current yield, 16


**D**

default probability, 234, 235
delta ($\Delta$), 123
    hedge, 123
    hedge limitation, 69
delta hedge, 66, 68, 122
derivative, 2, 225
    credit, 225
deterministic, 41
discount, 15, 16, 17, 35, 36, 50, 53,
    57, 60, 76, 98, 149, 150
    bond, 15, 17, 140
    factor, 35
    value, 93
discrete interest rate models, 158
discount currency, 215
diversification, 223
dividends, 2
drift, 61, 83, 84, 91
    zero, 83
dynamic hedging, 36, 37, 38, 68, 69,
    123, 124, 125, 130


**E**

equity, 2
Excel, 89
exercise price, 7
expected value, 31, 94
expiration, 3, 52, 59, 79, 115
expiration node, 99
export credit insurance, 224
extended Vasicek model, 186

**F**

face value, 15
fair game, 190, 191
Federal Reserve, 16, 221
fixed rate, 152
floating rate, 152
foreign currency, 225
foreign currency options, 211
foreign exchange, 207
forward contract, 3, 6, 209
forward interest rate, 18, 183
forward price, 213
forward rate, 183, 187, 209
    initial, 195
forward zero coupon price, 140
futures, 1, 12, 116
    Black–Scholes, 119
    price PDE, 118
FX, 207


**G**

gamma ($\Gamma$), 130, 131, 132
gamma hedge, 131
Garman–Kohlhagen, 211
GBM
    pricing model, 92, 94
    tree approximation, 100,
      102
geometric brownian motion,
    87
GER, 214, 215
GER call, 219
GER put, 219
global economy, 221
globalization, 222
guaranteed exchange rates
    (GER), 214, 215


**H**

hedging, 36–38, 122, 126, 152
    dynamic, 123
    limitations, 69
HJM bond price, 186
HJM model, 171, 183, 186

Ho-Lee, 197, 199
    model, 192
Hull–White term structure,
    184, 185

## I

implied volatility, 128
inflation, 16
initial yield curve, 177
interest rates, 16, 18, 19
    futures, 20
    models, 157, 158
    option, 137, 178, 183
    parity, 209
international risk, 221
Ito's lemma, 171, 212

## J

Jarrow-Lando-Turnbull (JLT), 234
JLT, 234
JLT example, 238

## K

Keynes parity, 209

## L

limited liability, 231
log normal model, 85
logarithmic plots, 86
Long Term Capital Management,
    221

## M

Maple, 91, 142
market price of risk, 173, 174, 178,
    180, 211
maturity period, 16

Merton, Robert, 230, 233
money market, 199
money market tree, 195, 197,
    199

## N

Nikkei futures, 215
no arbitrage, 2, 5, 6, 9, 28, 32, 41, 99, 150,
    151, 153, 166, 168, 172, 192

## O

open interest, 91
option, 1, 7, 9
    American, 52, 77
    cash or nothing, 114
    end in the money, 103
    exotic, 55, 59
    knockout, 55
    look back, 59
    stock or nothing, 115

## P

paradox, bond prices, 165
path probability, 60
political risk, 222, 224
    pricing, 228
portfolio, 4, 5, 26, 29, 93, 112, 172,
    173, 220
    differential, 120–121
    continuous time, 120
    replicating, 25, 29, 30, 31,
    67, 76, 93, 112
power call, 98
premium currency, 215
price spread, 37
primary market, 15
promissory note, 15
put, 25
    American, 10
    Black–Scholes, 97
    call parity, 97, 213
    European, 10, 97

put *(continued)*
  index option, 126
  option, 9


Q

quantos, 214


R

relative return, 61
replication, 5
repos, 18
risk premium, 229, 230, 233, 236, 237,
  238
risk-neutral probability, 32
Russian default, 234


S

secondary market, 15
self-financing, 112, 120
self-financing identity, 112
short rate, 183, 185
  adjusted, 198
  mean reverting, 178
short rate model, 171, 174, 185, 195
  simple, 174
short selling, 4
shorting, 4
Simpson's paradox, 165
spreadsheet, 71
  American, 77
  barrier option tree, 79
  calibration, 89
  call, 74
  formula, 72
  pricing, 76
  refined, 76
  stock, 75
standard normal, 83
stochastic, 41, 84
stochastic differential equation, 87
stock drift, 61
strike, 3
swap, 144
swap contract, 152–156

T

Taylor series, 110, 111
term structure, simple, 174
term structure models, 171
Thai baht, 221
Thai central bank, 177, 178
theta (Θ), 130
transition matrix, 235, 236
treasury bills, 21
tree
  calibration, 65, 87–89, 102
  scaling, 102
trend, 85


U

underwriters, 18
unfair game, 190
U.S. bond market, 15, 17


V

volatility, 119, 193, 194
  at maturity, 181
  bond, 181, 186
  deterministic, 186
  forward rate, 183, 188
  function, 173, 181
  short rate, 193, 195, 198
  smile, 129
  stock, 62
Vasicek, 179


Y

yield curve, 19, 138, 177, 179
yield spread, 229, 233
yield to maturity, 16, 17


Z

zero coupon bond, 15, 171, 231
zero coupon price, 150, 162–164, 171